AF615343

THE LIBRARY OF CONGRESS

Extra- and Intracellular Calcium and Phosphate Regulation:

From Basic Research to Clinical Medicine

Felix Bronner, Ph. D.
Professor Emeritus
Department of BioStructure and Function
University of Connecticut Health Center
Farmington, Connecticut

Meinrad Peterlik, M.D., Ph.D.
Professor and Head
Department of General and Experimental Pathology
University of Vienna Medical School
Vienna, Austria

CRC Press
Boca Raton Ann Arbor London Tokyo

Library of Congress Cataloging-in-Publication Data

Catalog record is available from the Library of Congress

This book represents information obtained from authentic and highly regarded sources. Reprinted material is quoted with permission, and sources are indicated. A wide variety of references are listed. Every reasonable effort has been made to give reliable data and information, but the authors and the publisher cannot assume responsibility for the validity of all materials or for the consequences of their use.

Direct all inquiries to CRC Press, Inc., 2000 Corporate Blvd., N.W., Boca Raton, Florida 33431.

International Standard Book Number 0-8493-0188-2

Printed in the United States of America 1 2 3 4 5 6 7 8 9 0

PREFACE

In recent years, biological research, including research in calcium and phosphate metabolism, has greatly accelerated in pace and provided answers to what until recently were only questions or speculations. The second Chemofux prize, awarded to a distinguished research contribution in the general area of bone and mineral metabolism published between 1988 and 1990, was shared by two investigators whose studies were characterized by the application of contemporary approaches to two important questions: The first of these — what is the nature of the gene that specifies the calcium-binding protein, a major component of transcellular calcium transport? — was answered by Monique Thomasset of Paris. The second — what are the details of the cellular events that take place in the course of bone resorption? — was analyzed by Anna Teti of Bari, Italy. As in the past, the two prize winners presented their papers at a scientific meeting in which members of the jury participated. As a result, it has been possible to structure an overview of current research and concepts, with the present book representing the outcome of this effort.

The current volume deals with four general topics: transport processes involving calcium and phosphate; genetic and functional analysis of the calbindins, molecules involved in calcium transport; hormonal regulation of the two minerals; and bone and bone diseases.

Chapter 1 analyzes the process by which calcium enters the cell, referring among other aspects to calcium channels, now recognized to be ubiquitous and of major importance in regulating calcium traffic. Chapter 2 describes two potential model cell systems for the study of cellular calcium transport. The human colon carcinoma cell line, CaCo-2, responds to stimulation by vitamin D, yet exhibits no increase in net calcium flux. It also does not express calbindin-9K. In contrast, cultured cells from the rabbit nephron, specifically the collecting duct, were found to transport calcium, to increase transport when stimulated by 1,25-dihydroxyvitamin D_3, the active form of vitamin D, and to contain calbindin 28K, the concentration of which increased upon stimulation by vitamin D. In Chapter 3, Caverzasio and colleagues discuss phosphate transport in osteoblastic cells, responsible for bone growth and modeling. They report that the sodium-dependent transport of phosphate by bone cells is enhanced by fluoride addition to the cell culture and raise the question whether this effect plays a role in the beneficial effect that has been attributed to fluoride treatment of osteoporosis.

Section II deals with the calbindins, relatively low molecular weight intracellular calcium-binding proteins thought to enhance the rate of intracellular diffusion of calcium, an enhancement necessary to account for the measured rate of transcellular calcium transport in the face of a very limited rate of self-diffusion of the calcium ion. Chapter 4 describes the prize winning work of Monique Thomasset, summarizing the structure of the gene for calbindin-9K (CaBP, Mr-9kD), its tissue-specific expression and its regulation by vitamin D. Chapter 5, by R. H. Wasserman, the discoverer of the calbindins, provides an overview of the role of the intestinal calbindins and of diseases that affect function or distribution of these molecules.

Section III, Hormonal Aspects, deals with a variety of regulatory phenomena. In Chapter 6, Martin elaborates on the functional role of the parathyroid hormone-related protein, for the purification, cloning and sequencing of which he had been awarded the first Chemofux prize. The protein is of significance because it appears to play a major role in the humoral hypercalcemia of malignancy and because many of its functional and biochemical characteristics are similar to those of the parathyroid hormone. Of particular interest is the possibility that this protein may play a role in the maintenance of the positive calcium gradient that exists between the fetal and maternal blood. Moreover, this control appears to be exerted by regions of the molecule that are not homologous with the parathyroid hormone molecule.

Chapter 7 deals with another new regulatory protein, chromogranin A, part of a family of molecules that seem to play a role in endocrine and neuroendocrine secretion. Chromogranin A in particular may function in the sorting and packaging of secretory proteins in endocrine cells, according to a scheme proposed by Cohn and Gorr.

Chapter 8 deals with a question that has puzzled endocrinologists, namely, why diseases of the thyroid gland also involve disturbances of calcium and mineral metabolism. Using findings from experiments with a variety of organ and cell cultures, the authors propose that thyroid hormones act in three ways, modulating the biosynthesis of 1,25-dihydroxyvitamin D_3 on the one hand, having a direct, vitamin D-like action on the cells of the principal organs involved in calcium and phosphate metabolism (gut, kidney and bone), on the other, and, finally, potentiating the effects of vitamin D on the cells of these three organs.

Section IV addresses a variety of questions related to bone and bone diseases. In Chapter 9 the other winner of the second Chemofux prize, Anna Teti, together with Alberta Zambonin Zallone, offers a detailed description of how the bone salt-dissolving activity of the osteoclast, the multinucleated cell responsible for bone destruction, may serve a feedback function that causes the osteoclast ultimately to stop its osteolytic activity on one site and move on to another. In Chapter 10, Stern deals with a topic of importance to the physician, namely, how drugs used for one purpose may deleteriously act on the skeletal system. For example, cyclosporin, a widely used immunosuppressant, has now been found to induce undesirable alterations in bone mass and turnover. Other drugs discussed are gonadotropin-releasing hormone, anti-estrogens, glucocorticoids, anticonvulsants, and thyroid hormones. Throughout *in vitro* and *in vivo* effects are compared and evaluated.

The discussion on secondary osteoporosis, Chapter 11, follows and expands on the description of the side effects of drugs on bone. Raisz emphasizes that osteoporosis that is secondary to such drug treatment may not result in the fractures typical of primary osteoporosis (postmenopausal or senile) and that the mechanisms by which secondary osteoporosis is brought about may vary widely, depending on which cells or metabolic pathways are targeted by the drugs. The final chapter provides a wide-ranging survey of the various drugs and therapeutic regimens used to treat metabolic bone disorders. Benefits and risks involved in treatment are discussed for virtually all drugs used, with emphasis on the fact that no effective agent is without side effects, and that side effects can differ even with the same agent and in different patients.

The title of this book may at first sight seem unduly ambitious, but we hope that our brief descriptions and the chapters themselves demonstrate how short the distance has become between experiment and therapy. We therefore hope the book will evoke interest from experimentalists and clinicians alike. At the same time, we wish to thank our coauthors for their contributions and the publishers for their help.

November, 1991

Felix Bronner
Farmington, Connecticut

Meinrad Peterlik
Vienna, Austria

THE EDITORS

Felix Bronner, Ph.D., is Professor Emeritus of BioStructure and Function and of Nutritional Sciences at The University of Connecticut. A graduate of the University of California at Berkeley and Davis and of the Massachusetts Institute of Technology, Dr. Bronner has worked in the general area of calcium metabolism for the past 40 years. He was among the first to study the kinetics of ^{45}Ca in humans, did detailed kinetic and balance studies in rats, and then proceeded to an analysis of calcium transport in the intestine and kidney. In recent years, he has returned to the problem of calcium homeostasis and its regulation. Throughout his career he has attempted to describe in quantitative fashion the events by which calcium moves in the body at the subcellular, cellular and organ levels, aiming for a conceptual synthesis.

Dr. Bronner has published over 80 research articles, has contributed some 50 chapters to texts and symposia volumes, and has edited more than 40 volumes of texts and symposia. He was founding editor of *Current Topics in Membranes and Transport* and first chair of the Gordon Research Conference on Bones and Teeth. He is currently on the editorial boards of the *Journal of Nutrition* and *The American Journal of Physiology*.

A member of numerous scientific societies, including fellowship in The American Association for the Advancement of Science, he was the 1974 recipient of the André Lichtwitz prize, awarded by the French National Institutes of Medical Research (INSERM) for outstanding work in calcium and phosphorus metabolism and the honoree of the 4th International Workshop on Calcium and Phosphate Transport Across BioMembranes, held in Lyon in 1989 under NATO sponsorship. He has organized and chaired national and international symposia and workshops and trained graduate students and many postdoctoral fellows.

Meinrad Peterlik, Ph.D., M.D., is Head of the Department of General and Experimental Pathology at the University of Vienna Medical School. In 1963 he received his Ph.D. in Chemistry and in 1972 his M.D. from the University of Vienna, Austria. In 1974–1975 he was granted a Max Kade Fellowship to pursue postdoctoral studies in Robert H. Wasserman's laboratory at Cornell University, Ithaca, NY. In 1978 he was promoted to Associate Professor (with tenure), and in 1983 he was appointed Full Professor of General and Experimental Pathology (Pathophysiology) at the University of Vienna.

Dr. Peterlik is a member of the American Institute of Nutrition, American Society for Bone and Mineral Research, and is currently President of the Austrian Society for Bone and Mineral Research. He serves as member of the Scientific Board of the European Calcified Tissue Society and of the Editorial Board of *Bone*. He is also the head of the International Jury for the Chemofux Prize for Bone and Mineral Research.

Dr. Peterlik (together with Dr. Felix Bronner) has organized a series of International Workshops on Calcium and Phosphate Transport Across Biomembranes in Vienna (1981, 1984 and 1987) and was Head of the Organizing Committee of the XXII European Symposium on Calcified Tissues (Vienna 1991).

Dr. Peterlik has published more than 70 original papers and has edited a textbook for medical students on pathophysiology (*Funktionelle Pathologie*). His current research interests include cellular actions and interactions of vitamin D and thyroid hormones, calcium and phosphate homeostasis, cytokines and bone turnover, and hormonal regulation of intestinal cell differentiation.

TABLE OF CONTENTS

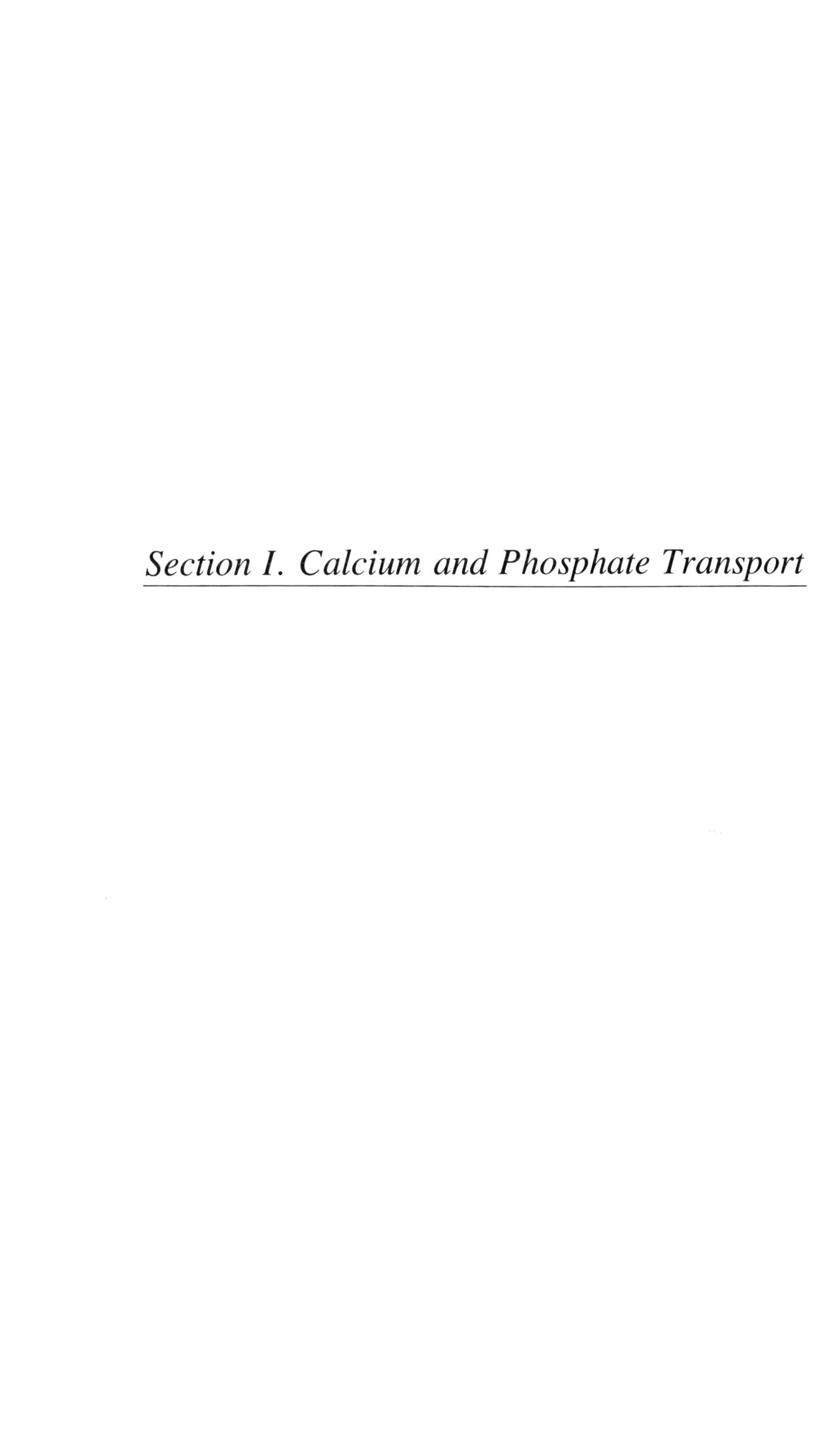

Section I. Calcium and Phosphate Transport

CALCIUM ENTRY AND CALCIUM CHANNELS

FELIX BRONNER
DEPARTMENT OF BIOSTRUCTURE AND FUNCTION
UNIVERSITY OF CONNECTICUT HEALTH CENTER
FARMINGTON, CT 06030

The resting free calcium ion concentration in mammalian cells tends to be less than 100 nM. Three systems are involved in regulating intracellular calcium: calcium entry, calcium extrusion and intracellular calcium binding sites of the organelles (mitochondria, endoplasmic reticulum, lysosomes, etc.). The binding capacity of the various organelles is so large that entry and extrusion may be considered the principal vectors of acute regulation. For example, when the calcium concentration of the buffer in which intestinal cells were bathed was increased from 1 to 3 mM, total calcium content of the cells increased in direct proportion.[1] Since the cells remained viable, their free calcium concentration may have increased, but most of the entering calcium must have been taken up by the many intracellular fixed binding sites.[1] Moreover, the calcium extrusion capacity of the Ca-ATPase is far in excess (perhaps 2-3 orders of magnitude) of the measured influx rate.[2] Inasmuch as the capacity to bind calcium that enters the cell is very large and its extrusion capacity is so much greater than the rate of entry, it is logical to consider entry as the rate-limiting step. As it turns out, however, the situation is not so simple, at least as far as the calcium-transporting cells are concerned.

The major regulator for transcellular calcium transport is vitamin D. Indeed, in the total absence of this compound - more specifically of the active metabolite, 1,25-dihydroxy-vitamin D_3, 1,25-$(OH)_2D_3$ - transcellular calcium transport grinds to a standstill. Yet calcium entry is only partly (30-40%) reduced in vitamin D deficiency.[3,4] Moreover, administration of 1,25-$(OH)_2D_3$ to vitamin D-deficient chicks [3] or rats [4] also enhances brush border membrane vesicle calcium uptake by only about 30-50 percent, whereas total transport increases in direct proportion to the vitamin D administered.[5]

The nature of vitamin D regulation of calcium entry may also differ from its effect on transcellular calcium transport. In the case of total transport it is becoming increasingly clear that 1,25-$(OH)_2D_3$ acts genomically, i.e. induces the biosynthesis of the cytosolic calcium-binding protein (calbindin D-9K, M_r=8.8 kD) via transcriptional, translational and possibly postranslational mechanisms.[6] As far as the entry mechanism is concerned, Kowarski and Schachter[7] have reported the existence of a vitamin D-dependent component of a brush-border-membrane-related particulate complex and have attributed to it a transport role. Miller and colleagues [8] have identified a membrane-bound, vitamin D-dependent calcium-binding protein from the brush-border that differs in several characteristics from the cytosolic calbindin. The brush-border calcium-binding protein identified by Miller et al[8] has an estimated molecular weight of 19kD and could therefore constitute the monomer of the protein identified by Kowarski and Schachter.[7] However, no further studies have appeared characterizing these molecules.

It is of course possible that 1,25-$(OH)_2D_3$ induces both the cytosolic and the brush-border calcium-binding proteins, but in that case the time required to induce the proteins should not differ significantly. In the case of calbindin D-9K, transcription takes some

4h[9], yet 1,25-$(OH)_2D_3$ has been shown[10] to act on brush-border membrane uptake of calcium in a substantially shorter period, provided the vitamin D metabolite was administered to the animal (and not in vitro). In osteosarcoma-derived osteoblast-like cells the addition of 1,25-$(OH)_2D_3$ to a cell culture produced an increase in calcium uptake in 1-2 min.[11] In similar cells the existence of calcium channels has been demonstrated[12] and a very rapid response leading to increased calcium influx was documented.[13] Although no reports on the existence of calcium channels in the brush border membranes of enterocytes have as yet appeared, these cells, like most others, are likely to have calcium channels. If so, the rapid response of such channels could account for the increase in calcium uptake observed in brush-border membrane preparations.[3,4] Furthermore, on a quantitative basis, the increase in calcium uptake by brush border membrane vesicles from treated animals matches the increase found in treated isolated osteosarcoma cells. However, this increase cannot, as stated above, account for most of the increase in transcellular calcium transport.

Calcium entry can be studied with the aid of brush-border membrane vesicles prepared from either the intestine or kidney. These vesicles are right-side out[14] and take up calcium in a time-dependent and concentration-dependent fashion, with an apparent K_m of 1.1 mM. Calcium that enters the vesicle becomes bound to the inner membrane. The membrane has two calcium-binding sites, a high-affinity site with modest binding capacity and a low affinity site with great binding capacity.[4]

Total calcium uptake by duodenal brush-border membrane vesicles is nearly two orders of magnitude too low as compared with the V_{max} of transcellular transport.[15] This may be the result of channel inactivation, as well as of difficulty in measuring an appropriate initial rate. Recent approaches[16] have improved measuring techniques, but current experimental values do not yet come close to the experimental transport value. If entry were to be a rate-limiting process, its basal rate would have to be either near the actual transport rate or be subject to modulation so as to increase it to the V_{max}. The intracellular self-diffusion rate of Ca^{2+} is two orders of magnitude smaller than the V_{max} for transcellular transport and can be increased in direct proportion to the tissue content of calbindin.[5] Calbindin D-9K content in turn varies directly with the quantity of 1,25-$(OH)_2D_3$ administered to vitamin D-deficient or vitamin D-replete animals.[5]

Since the available values on the V_{max} of calcium uptake by brush border membrane vesicles are too low for transcellular transport to take place at its measured rate and since modulation of the brush border calcium uptake is modest and insufficient to raise its V_{max} to that of transport, one cannot assert that calcium entry is the rate limiting step. One can, however, assert that the entry rate is significantly lower than is the exit or extrusion rate.[2] Moreover, if entry were to be blocked or diminished, this would necessarily limit the amount that is transported transcellularly.

The calcium channel blocker verapamil has been shown to decrease calcium uptake by about 55% in isolated brush border membrane vesicles.[4] Verapamil, administered to fasting rats intraluminally, diminished duodenal calcium absorption, evaluated by in situ intestinal sacs, when the calcium concentration of the instilled buffer was relatively low (2 mM),[17] i.e. when transcellular calcium transport predominated.[18] When however the instilled calcium concentration was high (50 mM), most of the calcium therefore being absorbed by the paracellular route,[15,18] verapamil was without effect.[17] These findings suggest that at least some calcium enters the transporting enterocyte via a dihydropyridine-sensitive channel. Calcium channels that are similar in characteristics to L-type channels[19] have also been found in renal cells.[20] In the latter, verapamil addition to the bath "reduced the fractional open time at negative but not at positive clamp potentials."[20] Nifedipine (100 mM) completely abolished channel activity which was restored to near maximal by addition of the agonist Bay K-644.[20] Saunders and Isaacson[20] therefore conclude that the channels they studied are specific for calcium. Hess and colleagues,[12] discussing the interaction of calcium with dihydropyridine-sensitive calcium channels, have concluded that the best documented mechanism for channel selectivity is one that involved differential affinities of permeant ions to intra-channel binding sites. In their view, at least one high-affinity calcium-binding site can be reached from either side of the channel, with access from the cytoplasmic side voltage-dependent. In such a situation, "externally located high-affinity Ca-binding sites with possible allosteric effects on L-type Ca channels..." would not be necessary.[21]

Petersen[22] has called attention to at least two other routes of calcium uptake by electrically non-excitable cells such as enterocytes. One is a non-specific or non-selective cation channel, the existence of which has been widely documented (see Ref. 20 for a listing of some). The other is a calcium-selective, voltage insensitive channel.[23] It is quite apparent (cf 19) that an increasing number of cation channels are being described. It will therefore be of interest to catalog the various calcium channels in cells that transport calcium and to quantitate how much of the entering calcium is handled by each.

The addition of 1,25-$(OH)_2$-D_3 to isolated clonal osteosarcoma cells produces a nearly instantaneous two-fold increase in calcium uptake that appears mediated by an inward calcium current, with maximal effects at negative test potentials.[13] The shifts in activation voltage dependence and the lengthening of the open times of single channels that may be attributed to the effect of low (<3nM) concentrations of 1,25-$(OH)_2D_3$ in the osteosarcoma cells could reflect, in the words of Caffrey and Farach-Carsen,[13] stabilization of channel open states. However, by what mechanism 1,25-$(OH)_2D_3$ acts on this high threshold L-type channel is not known. Thus whether intestinal cells respond like the osteosarcoma cells, whether they also possess these L-type channels and how the latter might respond to the vitamin D metabolite remain open questions. The 1,25-$(OH)_2D_3$ receptors that mediate the genomic response in intestinal and renal cells are cytosolic receptors[24] that, when charged with the hormone, travel to the nucleus. The nature of the receptor that permits the channel to alter its open time is as yet unknown.

Before concluding, attention should be called to the possibility that calcium entry into transporting cells may be via a carrier. This possibility was advanced by Holdworth[25] many years ago and has found indirect support in brush-border membrane studies by Schedl and collaborators.[26,27] The latter, like others[4,28], have analyzed brush-border membrane uptake as the sum of two processes, a linear and a saturating one, and have proposed that the saturating process is a mediated one, representing perhaps a carrier process. Others[4] have instead proposed that the saturating process represents binding to the inner aspect of the vesicle membrane. Indeed one can visualize three processes - rapid binding to the outer aspect of the vesicle membrane, transmembrane movement and binding to the inner aspect. There is general agreement that binding to the outer aspect is fastest, but no detailed kinetic analysis of calcium uptake by vesicles is available. As pointed out above, two classes of binding sites have been identified on brush-border membranes,[4] with high affinity and modest capacity and with relatively low affinity and high capacity, but their relative distribution on the inner and outer aspects of the brush border membranes is not known. Moreover, how these binding sites affect the kinetics of vesicular uptake is also not known. The possibility of a carrier cannot therefore be excluded. One can, however, argue that a cell equipped with two pump-leak systems - one involving facilitated diffusional uptake, the other an active extrusion pump - seems more complicated than necessary, particularly since the existence of a modulated calcium channel alone could account for other known features of cellular calcium uptake.

Finally one might ask questions concerning possible defects of calcium entry. In this connection it is worth recalling the words of Fleckenstein, the pioneer discoverer of calcium antagonists, that "...Ca channel blockade, which is producible in vitro only with huge overdoses,... can never occur in a living animal."[29] In other words, transcellular calcium flux is essential in a variety of cells, notably muscle, heart and calcium transporting cells, even if one were to ignore the intracellular calcium-signaling mechanisms that have evolved in most cells and that alone require some calcium entry. Calcium channels are therefore likely to be ancient, from an evolutionary viewpoint, and mutations that drastically affect channel function are probably lethal. Moreover, since calcium channels seem to alter calcium entry only about twofold, a defect that alters this ability may have effects that are relatively subtle and hard to detect. To be sure, in situations such as myocardial calcium overload, leading to necrosis, calcium antagonists are very useful,[29] but it is unclear whether the defect lies in excessive entry or defective extrusion capacity or some combination of both. Cloning of the genes of calcium channels, directed mutagenesis and the use of transgenic animals may shed light on the possible existence and nature of calcium channel diseases or defects.

REFERENCES

1. Bronner, F., Pansu, D., Bosshard, A., Lipton, J.H., Calcium uptake by isolated rat intestinal cells, J. Cell Physiol., 116, 322, 1983.
2. Lew, V.L., Permeability and permeabilization of red cells to calcium, in Intracellular Calcium Regulation, Bronner, F., ed., Wiley-Liss, New York, 1990, chap. 1.
3. Rasmussen, H., Fontaine, O., Max, E.E., Goodman, D.P., The effect of 1 (alpha)-hydroxyvitamin D_3 administration on calcium transport in chick intestine brush border membrane vesicles, J. Biol. Chem., 254, 2993, 1979.
4. Miller, A. III, Bronner, F., Calcium uptake in isolated brush-border vesicles from rat small intestine, Biochem. J., 196, 391, 1948.
5. Roche, C., Bellaton, C., Pansu, D., Miller, A. III, Bronner, F., Localization of vitamin D-dependent active Ca^{2+} transport in rat duodenum and relation to CaBP, Am. J. Physiol. 251, (Gastrointest. Liver Physiol. 14), G314, 1986.
6. Singh, R.P., Bronner, F., Vitamin D acts posttranscriptionally. In vitro studies with the vitamin D-dependent calcium-binding protein of rat duodenum, in Calcium Binding Proteins: Structure and Function., Siegel, F.L., Carafoli, E., Kretsinger, R.H., MacLennan, D.H., Wasserman, R.H., eds., Elsevier-North Holland, New York, 1980, 379.
7. Kowarski, S., Schachter, D., Intestinal membrane calcium-binding protein: Vitamin D-dependent membrane component of the intestinal calcium transport mechanism, J. Biol. Chem., 255, 10834, 1980.
8. Miller, A. III, Ueng, T.H., Bronner, F., Isolation of a vitamin D-dependent, calcium-binding protein from brush borders of rat duodenal mucosa, FEBS Letters, 103, 319, 1979.
9. Warembourg, M., Perret, C., Thomassset, M., Distribution of vitamin D-dependent calcium-binding protein messenger RNA in rat placenta and duodenum, Endocrinology, 119, 176, 1986.
10. Matsumoto, T., Fontaine, O., Rasmussen, H., Effect of 1,25-dihydroxyvitamin D_3 on phospholipid metabolism in chick duodenal mucosal cell. Relationship to its mechanism of action, J. Biol. Chem., 256, 3354, 1981.
11. Lieberherr, M., Effects of vitamin D_3 metabolites on cytosolic free calcium in confluent mouse osteoblasts, J. Biol Chem., 262, 13168, 1987.
12. Yagamuchi, D.T., Hahn, T.J., Iida-Klein, A., Kleeman, C.R., Muallem, S., Parathyroid hormone-activated calcium channel in an osteoblast-like clonal osteosarcoma cell line. cAMP dependent and cAMP-independent channels, J. Biol Chem., 262, 7711, 1987.
13. Caffrey, J.M., Farach-Carson, M.C., Vitamin D_3 metabolites modulate dihydropyridine-sensitive calcium currents in clonal rat osteosarcoma cells, J. Biol. Chem. 264, 20265, 1989.
14. Haase, W., Schafer, A., Murer, H., Kinne, R., Studies on the orientation of brush-border membrane vesicles, Biochem. J., 172, 57, 1978.
15. Bronner, F., Pansu, D., Stein, W.D., An analysis of intestinal calcium transport across the rat intestine, Am. J. Physiol., 250, (Gastrointest. Liver Physiol. 13), G561, 1986.
16. Ghijsen, W.E.J.M., Ganguli, U., Stange, G., Gmaj, P., Murer, H., Calcium uptake into rat small intestinal brush border membrane vesicles: characterization of transmembrane calcium transport at short initial incubation times, Cell Calcium, 8, 157, 1987.
17. Fox, J., Green, D.T., Direct effects of calcium channel blockers on duodenal calcium transport in vivo, Europ. J. Pharmacol., 129, 159, 1986.

18. Pansu, D., Bellaton, C., Bronner, F., The effect of calcium intake on the saturable and non-saturable components of duodenal calcium transport, Am. J. Physiol., 240, (Gastrointest. Liver Physiol. 3), G32, 1981.
19. Miller, R.J., Fox, A.P., Voltage-sensitive calcium channels, in Intracellular Calcium Regulation, Bronner, F., ed., Wiley-Liss, New York, 1990, 97.
20. Saunders, J.C.J., Isaacson, L.C., Patch clamp study of Ca channels in isolated renal tubule segments, in Calcium Transport and Intracellular Calcium Homeostasis, Pansu, D., Bronner, F., eds. NATO/ASI series H, vol. 48, Springer, Berlin, 1990, 27.
21. Hess, P., Prod'hom, B., Pietrobon, D., Mechanisms of interaction of permeant ions and protons with dihydroxypyridine-sensitive calcium channels, Ann. N.Y. Acad. Sci., 560, 80, 1989.
22. Petersen, O.H., Regulation of calcium entry in cells that do not fire action potentials, in Intracellular Calcium Regulation, Bronner, F., ed., Wiley-Liss, New York, 1990, 77.
23. Zschauer, A., van Breeman, C., Buhler, F.R., Nelson, M.T., Calcium channels in thrombin-activated human platelet membrane, Nature, 334, 703, 1988.
24. McDonnell, D.P., Mangelsdorf, D.J., Pike, J.W., Haussler, M.R., O'Malley, B. W., Molecular cloning of complementary DNA encoding the avian receptor for vitamin D, Science, 235, 1214, 1987.
25. Holdsworth, E.S., The effect of vitamin D on enzyme activities in the mucosal cells of the chick small intestine, J. Membrane Biol., 3, 43, 1970.
26. Schedl, H.P., Wilson, H.D., Calcium uptake by intestinal brush border membrane vesicles. Comparison with in vivo calcium transport, J. Clin. Invest. 76, 1871, 1985.
27. Wilson, H.D., Schedl, H.P., Christensen, K., Calcium uptake by brush-border membrane vesicles from the rat intestine, Am J. Physiol. 257, (Renal Fluid Electrolyte Physiol. 26), F446, 1989.
28. Takito, J., Shinki, T., Sasaki, T., Suda, T., Calcium uptake by brush-border and basolateral membrane vesicles in chick duodenum, Am. J. Physiol., 258, (Gastrointest. Liver Physiol. 21), G16, 1990.
29. Fleckenstein, A., History of calcium antagonists, Circul. Res. 52, Supplement 1, I-3, 1983.

TRANSCELLULAR CALCIUM TRANSPORT ACROSS CULTURED INTESTINAL AND RENAL CELLS

R.J.M.Bindels, J.A.H. Timmermans, A. Hartog and C.H. van Os
Department of Physiology, University of Nijmegen
P.O.Box 9101, 6500 HB Nijmegen (The Netherlands)

SUMMARY

The human colon carcinoma cell line CaCo-2 and a primary culture of rabbit kidney collecting duct system were tested whether they provide good model systems for studying active transcellular Ca^{2+} transport. Confluent monolayers of CaCo-2 cells grown on permeable supports were placed in modified Ussing chambers. J_{MS} and J_{SM} fluxes of Ca^{2+} increased significantly after 48 h preincubation with 10^{-7}M $1,25(OH)_2D_3$, without a significant effect on the net Ca^{2+} flux. $^{45}Ca^{2+}$ uptake across the luminal membrane of CaCo-2 cells grown on plastic dishes was also responsive to $1,25(OH)_2D_3$. Calbindin-D_{9K} could not be detected in CaCo-2 cells with a Mab against human Calbindin-D_{9K}.

Rabbit collecting duct system cells were isolated by immunodissection and subsequently cultured on permeable supports. In this way 80% of Calbindin-D_{28K} containing cells were isolated. After 6 days in culture, confluent monolayers transported 83±6 nmol Ca^{2+}/h.cm^2 from the apical to the basal solution. This Ca^{2+} transport was largely dependent on serosal Na^+. Exposure to 0.1 μM $1,25(OH)_2D_3$ for 48 h increased transcellular Ca^{2+} transport by 53% and PTH stimulated net Ca^{2+} transport by 25%. The cells in primary culture contain Calbindin-D_{28K}, which content is increased after 48 h exposure to $1,25(OH)_2D_3$.

In conclusion, CaCo-2 cells are responsive to $1,25(OH)_2D_3$, but do not express net Ca^{2+} transport. The primary culture of renal collecting duct system proved to be a promising model system to study regulation of active transcellular Ca^{2+} transport.

INTRODUCTION

In intestinal and renal epithelia there are two routes available for Ca^{2+} transport. Passive Ca^{2+} movements in both directions occur across the paracellular route, while active Ca^{2+} transport is exclusively transcellular from lumen to blood side. The transcellular route consists of a passive influx, diffusion through the cytosol and active extrusion mediated by the plasma membrane type of Ca^{2+}-ATPase and/or a Na^+/Ca^{2+} exchanger[1]. In addition, a variety of ion transport processes in intestinal and renal cells are regulated by the level of intracellular free Ca^{2+} concentration, $[Ca^{2+}]_i$ [2,3]. Hence, receptor-operated Ca^{2+} signalling in these particular cells is continuously challenged by large and variable rates of transcellular Ca^{2+} transport. Since there is no evidence for an effect of the vitamin D status on basal electrolyte transport rates, it seems that transcellular Ca^{2+} transport does not interfere with Ca^{2+} signalling. How intestinal and renal cells make the distinction between Ca^{2+} in transit and Ca^{2+} as intracellular messenger is unclear. In order to be able to study in detail hormone-regulated Ca^{2+} fluxes and receptor-operated Ca^{2+} signalling in these cells, it would be of great help when cultured cells were available which express transcellular Ca^{2+} transport activity.

This contribution reports on cultured intestinal and renal epithelial cells which can be used as model systems in studying regulation of transcellular Ca^{2+} transport.

METHODS AND MATERIALS

CELL CULTURES

The human colon adenocarcinoma cell line CaCo-2 was obtained from the American Type Tissue Collection (passage 15) and studied between the 16th and 30th passage. CaCo-2 cells were routinely maintained at 37 °C in a humidified 5% CO_2-95% air atmosphere in Dulbecco's modified Eagle's medium, supplemented with 5% fetal bovine serum, 100 μg/ml penicillin, 100 μg/ml streptomycin and a mixture of non-essential amino acids. The cells were fed three times a week and were trypsinized weekly with PBS solution containing 0.1% (w/v) trypsin and 1 mM EDTA.

Renal cells from the connecting tubule and cortical collecting duct were isolated by immunodissection as described in detail by Bindels et al.[4]. Briefly, four New Zealand white rabbits ($\approx$ 0.5 kg weight) were killed by cervical dislocation and bled. The kidneys were removed and placed in Krebs-Henseleit buffer, KHB (composition in mM: 128 NaCl; 5 KCl; 1 $MgSO_4$; 1 $CaCl_2$; 2 NaH_2PO_4; 10 glucose; 10 Na-acetate; 4 l-lactate; 1 l-alanine; 20 HEPES/Tris, pH 7.40). All procedures were performed under sterile conditions. A thin layer of superficial cortical tissue was carefully dissected, pooled and finely diced with a razor blade. The minced tissue was incubated in a shaker bath at 37 °C for 30 min in 10 ml 0.2% w/v collagenase A, 0.1% w/v hyaluronidase in KHB. The sample was centrifuged at 200 x g for 5 min. The tissue pellet was resuspended in 20 ml 10 mM EDTA in KHB and incubated at 37 °C for 10 min in a shaker bath. At 5 min intervals, the sample was dispersed by pipetting. Finally, 40 ml KHB was added to the dispersed cell suspension, sieved through nylon filters (100 & 45 μm) and centrifuged at 200 x g for 5 min. The resulting cell pellet was resuspended in 10 ml hybridoma medium containing Mab R2G9. The incubation was continued for 60 min on ice with occasional shaking. The cells were collected by centrifugation (200 x g for 5 min), washed three times with KHB containing 5% v/v FCS and resuspended in the same solution. Subsequently, 5 ml of this suspension was added to each of six goat anti-mouse IgM-coated bacterial dishes. The cells remained on these dishes for 15 min at room temperature. Each dish was washed carefully for 6 times with 5 ml KHB containing 5% v/v FCS. The adherent cells were scraped off the dishes and collected by centrifugation (200 x g for 5 min).

Immunodissected cells of the collecting duct system were seeded at a high density of $1.5.10^6$ cells/cm^2 on 0.3 cm^2 permeable filters (Costar, Badhoevedorp, NL) coated with rat tail collagen. The filters were placed in 24-well plates (Costar, Badhoevedorp, NL) and incubated with 100 μl medium at the mucosal side (inside of the filter cup) and with 600 μl medium at the serosal side (outside of the filter cup). The culture medium was a 1:1 mixture of DME/F12 medium (Gibco, Breda, NL) supplemented with 5% v/v decomplemented FCS, 50 μg/ml gentamicin (Schering, Kenilworth, NJ), a mixture of non-essential amino acids (Flow, Irvine, UK) 5 μg/ml insulin, 5 μg/ml transferrin, 50 nM hydrocortisone, 70 ng/ml PGE_1, 50 nM Na_2SeO_3 and 5 pM triiodothyronine, equilibrated with 5% CO_2-95% air at 37 °C. Medium was changed 24 h after cell seeding and successively every 48 h. All experiments were performed between 6-9 days after seeding the cells on permeable filters.

<u>Ca^{2+} INFLUX MEASUREMENTS</u>

All experiments were performed with CaCo-2 monolayers, grown on 12-wells plates. Immediately before the Ca^{2+} uptake experiments, the cells were washed with buffer containing 140 mM NaCl, 5.4 mM KCl, 1.2 mM $MgSO_4$, 0.1 or 0.7 mM $CaCl_2$, 10 mM HEPES-Tris (pH 7.4, 37 °C). $^{45}Ca^{2+}$ uptake was initiated by adding the same buffer containing 2 μCi/ml $^{45}Ca^{2+}$. The $^{45}Ca^{2+}$ uptake was stopped by addition of ice-cold 150 mM NaCl, 2.5 mM $CaCl_2$, 20 mM HEPES-Tris (pH 7.4). After several washings with this buffer, an aliquot of the cells was taken and counted for radioactivity.

<u>TRANSCELLULAR Ca^{2+} FLUX MEASUREMENT</u>

Transepithelial Ca^{2+} fluxes across CaCo-2 were measured in Ussing-type flux chambers. The 0.3 cm^2 filters were placed between two half-chambers and bathed with KHB buffers and continuously gassed. $^{45}Ca^{2+}$ fluxes were measured by adding $^{45}Ca^{2+}$ to

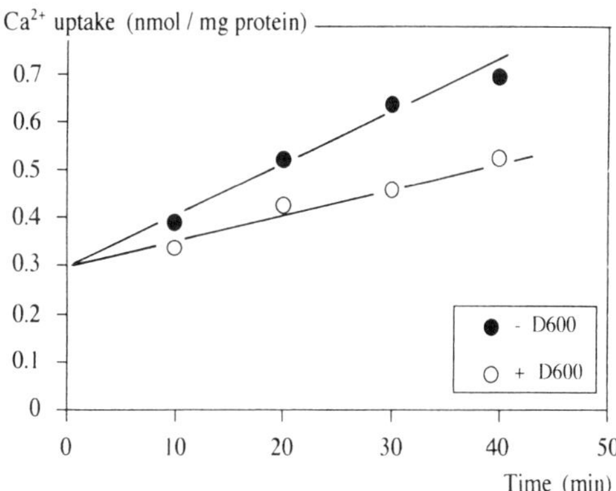

<u>Figure 1</u>. Time course of $^{45}Ca^{2+}$ uptake in confluent CaCo-2 cells in the presence of vehicle (O) or 1 mM D600 (O).

either the mucosal or serosal solution. Hundred μl samples were taken at 10 min intervals from the "cold" side. Samples were counted for radioactivity.Transepithelial Ca^{2+} flux across cultured renal monolayers was assessed by two different methods. Firstly, the filter cups were washed twice with culture medium containing 1 mM Ca^{2+}, and 4 μCi/ml $^{45}Ca^{2+}$ was added to either the mucosal or serosal compartment. At 10 min intervals for 80 min, 5 μl samples were taken from the non-labelled side and counted for radioactivity. Net Ca^{2+} flux (J_{net}) was calculated as the difference between mucosal-to-serosal flux (J_{MS}) and serosal-to-mucosal flux (J_{SM}). Secondly, the filter cups were washed twice with culture medium containing 1 mM Ca^{2+}. After a certain time, 20 μl samples were taken from the mucosal and serosal compartment, and the Ca^{2+} concentration in the samples was determined (Lancer, St.Louis, MO).

Hormone-sensitive Ca^{2+} fluxes were determined as follows. The monolayers were incubated at the serosal side with 10^{-7}M bPTH(1-34) in culture medium. After 1 h, samples were taken from both sides of the filter to determine the Ca^{2+} concentration, net Ca^{2+} absorption being calculated in $nmol.h^{-1}.cm^{-2}$. The effect of $1,25(OH)_2D_3$ was investigated by preincubating the cells with 10^{-7}M $1,25(OH)_2D_3$ on the serosal side for 48 h.

ENZYMES AND HORMONES

Collagenase A and hyaluronidase were obtained from Boehringer (Mannheim, FRG); $1,25(OH)_2D_3$ was kindly provided by Duphar (Weesp, NL). All other chemicals and conjugated antibodies were from Sigma (St.Louis, MO). bPTH(1-34) was dissolved in 5 mM acetic acid containing 0.1% w/v albumine. $1,25(OH)_2D_3$ was dissolved in ethanol. The effect of the various hormones tested was always evaluated by comparison with solvent treatment alone, not exceeding 0.1% v/v/ solvent in the incubate.

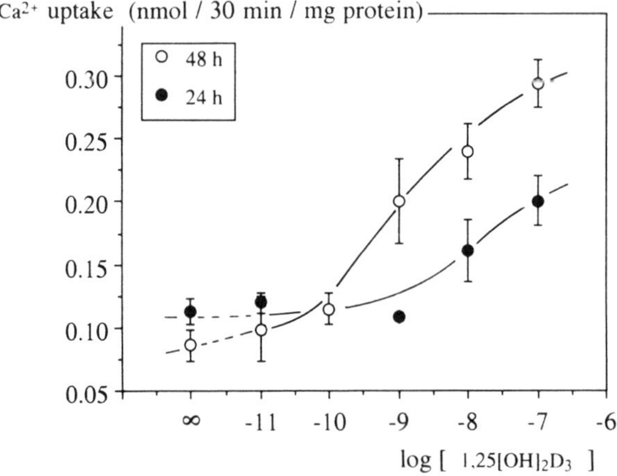

Figure 2. Dose-response curve of $1,25(OH)_2D_3$ on ^{45}Ca-uptake across the luminal membrane (n=4). CaCo-2 cells are grown on plastic 12 wells plates. ^{45}Ca-uptake is measured during 30 min at 37 •C.

STATISTICAL ANALYSIS

All reported data are expressed as the mean ± SE. Statistical differences between the mean values were determined by analysis of variance. Unpaired Student's t-test was used to determine the statistical difference between two independent groups.

RESULTS

Ca^{2+} TRANSPORT PROPERTIES OF CaCO-2 MONOLAYERS

CaCo-2 cells grown on 12-wells plates and on permeable membranes were confluent at day 14. The layers possessed many domes, characteristic of functionally polarized, transporting epithelial cells. The electrical resistance of the monolayers grown on permeable membranes ranged between 215 to 290 Ω.cm^2. The $^{45}Ca^{2+}$ influx into CaCo-2 cells was linear up to 40 min. (Figure 1). The Ca^{2+} entry blocker D600 (10^{-4} M) inhibited 50% of tjis influx, but was without effect on the Ca^{2+}-binding to the cells (intercept with Y-axis.). Therefore, the effect of 1,25(OH)$_2$D$_3$ on Ca^{2+} influx was determined after 30 min incubation with ^{45}Ca. The difference between the absence and presence of D600 was taken as influx in order to correct for aspecific binding. The effect of 1,25(OH)$_2$D$_3$ on Ca^{2+} influx was assessed by incubating CaCo-2 cultures (day 14) for 24 h and for 48 h in the absence or presence of 10^{-11} to 10^{-7}M 1,25(OH)$_2$D$_3$. Increasing the 1,25(OH)$_2$D$_3$ concentration, an dose-dependent increase in Ca^{2+} influx was observed (Figure 2). Incubation of the cells with 1,25(OH)$_2$D$_3$ for 48 h instead of 24 h resulted in higher Ca^{2+} influxes.

The effect of 1,25(OH)$_2$D$_3$ on transepithelial Ca^{2+} transport in confluent CaCo-2 cells (day 14) grown on permeable membranes was assessed after 48 h incubation with 10^{-7}M 1,25(OH)$_2$D$_3$. Both J_{MS} and J_{SM} fluxes were significantly increased by 1,25(OH)$_2$D$_3$ (Figure 3). However, there was no effect of the hormone on the net flux. We conclude

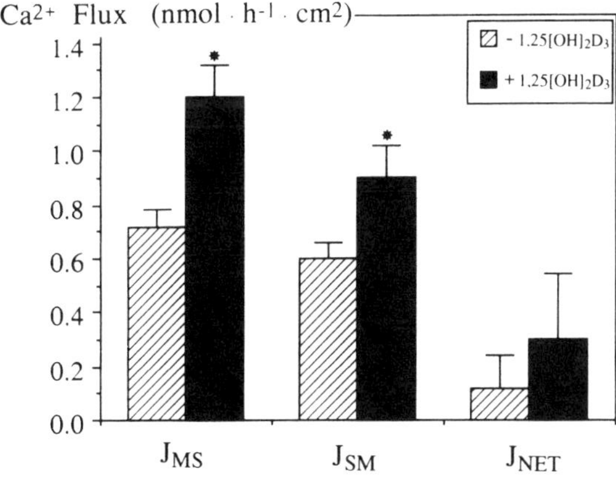

Figure 3. Unidirectional ^{45}Ca-fluxes across confluent monolayers of CaCo-2 in Ussing chambers (n=6). fluxes are measured at 37 °C during 1 h.

therefore, that $1,25(OH)_2D_3$ does not affect transcellular Ca^{2+} transport in confluent CaCo-2 cells. The electrical resistance of CaCo-2 cells monolayers were significantly higher in the presence of $1,25(OH)_2D_3$; 274 ± 13 Ω.cm² (n=5) versus 226 ± 10 Ω.cm² (n=5) (P < 0.001) in the absence of the hormone.

Ca^{2+} TRANSPORT PROPERTIES OF PRIMARY CULTURES OF COLLECTING DUCT SYSTEM

Rabbit cells of the collecting duct system were isolated from superficial cortex of rabbit kidney by immunodissection with Mab R2G9 and subsequently cultured. In this way, a relatively large number (5.10^6 cells/kidney) of highly purified collecting tubule cells were isolated. When grown on permeable filters, cultures reached confluence within 4 days after seeding, as evaluated by the asymmetric distribution of electrolytes on both sides of the filter, and by the development of a lumen negative P.D. across the monolayer of -24±3 mV, concomitantly with a transepithelial resistance of 284 ± 19 Ω.cm².

Isotopic Ca^{2+} fluxes from the apical to basal compartment were linear with time for 1.5 h, whereas no $^{45}Ca^{2+}$ flux in the opposite direction could be detected. A net $^{45}Ca^{2+}$ flux of 121 ± 13 $nmol.h^{-1}.cm^{-2}$ was observed (Figure 4). Net Ca^{2+} transport could also be estimated from the decrease in apical Ca^{2+} concentration with time. In this way a net Ca^{2+} absorption of 91 ± 15 $nmol.h^{-1}.cm^{-2}$ was calculated, not significantly different ($p > 0.2$) from radioisotopic flux measurements[4]. Ca^{2+} absorption across the monolayers was strongly dependent on serosal Na^+, since complete replacement of Na^+ in the serosal compartment by N-methylglucamine resulted in a 60% reduction of net Ca^{2+} transport (Table 1).

The effect of PTH and $1,25(OH)_2D_3$ on Ca^{2+} absorption across the monolayers was examined and the results are given in Table 2. After a 48 h preincubation period with 10^{-7}M $1,25(OH)_2D_3$, Ca^{2+} transport was significantly increased by 53% ($p < 0.05$). On the other hand, PTH had an immediate effect, since 1 h following addition of 10^{-7}M bPTH(1-34) to

TABLE 1.
Na^+ dependence of transcellular Ca^{2+} transport in collecting duct system in primary culture

	J_{NET} (nmol Ca^{2+}/h.cm²)	
Control	99 ± 6	(6)
$[Na^+]_{Basal}$=0	30 ± 5*	(6)
Ouabain	58 ± 1*	(3)

Values are mean ± SEM. Number of experiments between parenthesis. Ouabain (1 mM) was applied only to the basal compartment. *, significantly different fromcontrol ($p < 0.05$).

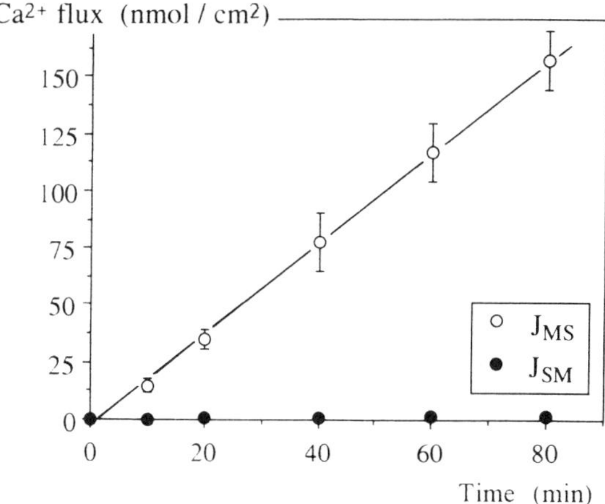

Figure 4. Unidirectional ^{45}Ca-fluxes across confluent connecting tubule cells in primary culture (n=4). J_{MS} equals J_{NET}.

TABLE 2.
Hormone-dependence of Ca^{2+} transport in collecting duct system in primary culture

	Ca^{2+} absorption (% of control)
$1,25(OH)_2D_3$	153 ± 12*
bPTH(1-34)	124 ± 8*
$1,25(OH)_2D_3$ + bPTH(1-34)	159 ± 10*

Values are mean ± SE of 10 experiments. In each experiment the amount of Ca^{2+} absorbed within the control period is set at 100% and averaged: 83±6 nmol/h.cm². In all experiments, hormones were applied only to the serosal side at a concentration of 10^{-7}M. Preincubations with $1,25(OH)_2D_3$ (10^{-7}M, serosal side) were performed for 48 h. *, significantly different from control ($p < 0.05$).

the basal compartment, Ca^{2+} transport was significantly increased by 24% ($p < 0.05$). Furthermore, PTH had no effect on Ca^{2+} absorption when the monolayers were preincubated for 48 h with $1,25(OH)_2D_3$. The actions of both hormones were, therefore, neither additive nor potentiating. In addition, a dose-response curve for $1,25(OH)_2D_3$ was determined. Ca^{2+} transport across the monolayers increased, when the concentration of $1,25(OH)_2D_3$ was raised from 10^{-10} to 10^{-7}M, with 10^{-10}M already increasing Ca^{2+} absorption significantly ($p < 0.05$), and a maximal effect was achieved at 10^{-8}M (Figure 5).

Rabbit connecting tubules and cortical collecting ducts normally contain considerable amounts of calbindin-D_{28K}. The presence of this Ca^{2+}-binding protein in the cultured cells was, therefore, examined, as well as the influence of $1,25(OH)_2D_3$. Western blot analysis of a cytosolic fraction of rabbit connecting tubules in culture demonstrated a band at 28 kD, which co-migrated with purified rabbit calbindin-D_{28K}[4]. To compare the calbindin-D_{28K}-content of the cultured cells, an inhibition ELISA system was used[4]. Calbindin-D_{28K}-content of the cells cultured in the presence of $1,25(OH)_2D_3$ was significantly increased compared to the control value (Table 3).

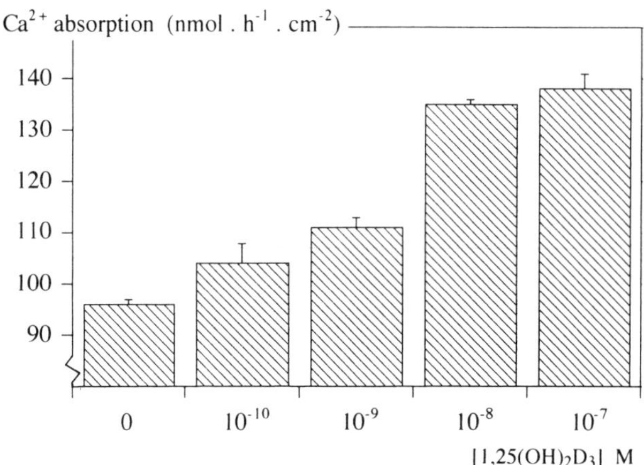

Figure 5. Effect of $1,25(OH)_2D_3$ on Ca^{2+} transport in primary cultures of rabbit collecting duct system. Preincubations with $1,25(OH)_2D_3$ (basal side) were performed for 48 h. Significant differences from control value (treated with vehicle above) were observed for all concentrations of $1,25(OH)_2D_3$ tested ($p < 0.05$). Values are mean ± SE of 4 experiments.

TABLE 3.
Effect of $1,25(OH)_2D_3$ on calbindin-D_{28K} content of collecting duct system in primary culture

	Calbindin-D_{28K} (μg/mg protein)
Control	2.0 ± 0.6
$1.25(OH)_2D_3$	9.4 ± 2.2*

Values are mean ± SE of 11 experiments. Preincubations with $1,25(OH)_2D_3$ (10^{-7}M, basal side) were performed for 48 h prior to the experiment. *, significantly different from control ($p < 0.01$).

DISCUSSION

Our study with CaCo-2 cells after 14 days in culture demonstrated an increase in $^{45}Ca^{2+}$ uptake after 24 h preincubation with $1,25(OH)_2D_3$. We also studied the effect of $1,25(OH)_2D_3$ on transepithelial Ca^{2+} transport in CaCo-2 monolayers. Although there was a significant increase in the J_{MS} as well as J_{SM} flux after preincubation with 10^{-7}M $1,25(OH)_2D_3$, there was no statistically significant effect on net Ca^{2+} absorption. This means that there is an effect of $1,25(OH)_2D_3$ on the transcellular route and also on the paracellular route leading to an increased J_{SM} flux. In a recent study, stimulation of the J_{MS} Ca^{2+} flux in CaCo-2 cells by $1,25(OH)_2D_3$ was reported[6]. However, in that study no J_{SM} flux was determined, hence it is unclear whether these authors were able to demonstrate significant net Ca^{2+} fluxes across CaCo-2 monolayers. The transepithelial Ca^{2+} flux across $1,25(OH)_2D_3$ treated CaCo-2 cells is 50-fold lower than found in cultured rabbit collecting tutular cells. This might be due to the high concentration of calbindin-D_{28K} present in collecting tubular cells in primary culture. Calbindin-D_{28K} might act as a soluble Ca^{2+} shuttle between the apical and the basal side of the cell, thereby facilitating transcellular diffusion of Ca^{2+}. We were unable to demonstrate the presence of Calbindin-D_{9K} in CaCo-2 cells, despite the fact that we produced a monoclonal antibody against purified human intestinal calbindin-D_{9K}. Although we found an increase in Ca^{2+} influx into CaCo-2 cells by $1,25(OH)_2D_3$, the rate limiting step of transcellular Ca^{2+} transport is most likely diffusion of Ca^{2+} through the cytosol[7], hence an effect on net Ca^{2+} absorption should not be anticipated in the absence of calbindin.

A primary culture of rabbit kidney collecting duct system containing high amounts of calbindin-D_{28K} was established, which proved to be a better system for studying cellular mechanisms of active Ca^{2+} transport than CaCo-2. These renal cells were isolated using an immunoaffinity procedure with monoclonal antibody R2G9, which recognizes a cell surface antigen specific for connecting tubules and collecting ducts[4]. In this way, tubular cells representing 80% of calbindin-D_{28K}-containing cells were isolated from the superficial rabbit kidney cortex and subsequently cultured on permeable filters. The resultant culture system constituted a population of cells from both connecting tubules and cortical collecting ducts and retained many characteristics of the collecting duct system[4].

Calbindin-D_{28K} was expressed by the cultured cells, corresponding to 'in vivo' studies where calbindin-D_{28K} has been located in distal convoluted tubule, connecting tubule and cortical collecting duct[5]. Using a competitive ELISA, the calbindin-D_{28K} content of the cultured cells was 2 μg/mg protein, which is in the same order of magnitude to that reported for a rat kidney cortex cytosolic fraction[8]. Since 1 mg protein contains 2.10^6 cells and assuming a radius of 5 μm for cultured cells, the average calbindin-D_{28K} concentration in the cultured cells amounts to at least 70 μM. According to Feher et al., the calbindin concentration in the cytosol should reach values of 50 μM or higher in order to facilitate cytosolic Ca^{2+} diffusion significantly[9]. This means that in cultured cells of the collecting duct system, calbindin concentration is sufficiently high to play a significant role in cytosolic Ca^{2+} movement.

This study also demonstrates that cells of the collecting duct system in primary culture are capable of transporting Ca^{2+} via a transcellular route. Net basal Ca^{2+} absorption across the monolayers ranges between 15 and 20 $pmol.min^{-1}.mm^{-2}$. These values agree well with data from 'in vivo' and 'in vitro' studies in the distal nephron. Using micropuncture techniques in rat distal tubule, Costanzo and Windhager measured Ca^{2+} absorption of 32 $pmol.min^{-1}.mm^{-2}$ [10]. In isolated perfused rabbit connecting tubules, Imai et al. report values around 3 $pmol.min^{-1}.mm^{-1}$ [11], and assuring a tubular radius of 10 μm, this amounts to 48 $pmol.min^{-1}.mm^{2}$. confluent monolayers of cells from the collecting duct system exhibit no significant J_{SM} Ca^{2+} flux which agrees with 'in vivo' studies in rat distal tubule[10] and which emphasizes the "tightness" of the tight junctional complex in this part of the nephron.

The connecting tubule, at least in the rabbit, is one of the major sites where PTH increases Ca^{2+} reabsorption[11]. Accordingly, the present study demonstrates a PTH-induced stimulation of Ca^{2+} absorption in rabbit collecting duct system in primary culture. However, the PTH effect was small compared to data from microperfusion studies, where a stimulation by PTH of up to 400% of basal rates was observed. This may be due to the heterogeneous nature of the primary culture, in which the connecting tubular cell might be a minority.

In contrast to PTH, little information is available on direct effect of vitamin D on Ca^{2+} transport in the kidney[1, 12, 13]. The present study demonstrates that $1,25(OH)_2D_3$ not only stimulates transcellular Ca^{2+} absorption but also increases calbindin-D_{28K} content in primary cultures from rabbit collecting duct system. Our findings are in line with studies performed in the mammalian intestine, where $1,25(OH)_2D_3$ also stimulates Ca^{2+} absorption and increases the cellular content of calbindin-D_{9K} [14]. Further studies will be performed to investigate the relationship between calbindin-D_{28K} content and $1,25(OH)_2D_3$-stimulated Ca^{2+} absorption in cortical collecting duct system.

Transcellular Ca^{2+} absorption across epithelia involves passive influx across the luminal membrane, followed by active extrusion of Ca^{2+} at the opposing basolateral plasma membrane[1]. In the luminal membrane, Ca^{2+}-entry mechanisms are largely unknown [1]. Two mechanisms for Ca^{2+} extrusion have been demonstrated in basolateral membrane, a Ca^{2+}-ATPase and a Na^+/Ca^{2+} exchanger[1]. Recent reports on intracellular Ca^{2+} rabbit connecting tubules [15] and cortical collecting ducts [16], and on transtubular Ca^{2+} transport in rabbit connecting tubule [11], suggest, that a Na^+/Ca^{2+} exchanger in the basolateral membrane may participate in the transcellular Ca^{2+} absorption. This is now confirmed in cultured cells, since transcellular Ca^{2+} absorption largely depends on serosal Na^+.

We conclude, that a primary culture of rabbit collecting duct system obtained by immunodissection retained many of the original transport characteristics. Although the cultured cells originated mainly from rabbit connecting and collecting tubules, several different cell types may still be present in these monolayers. The distal nephron is a complex segment consisting of loops of Henle, distal convoluted tubules, connecting tubules and collecting ducts, and most segments contain at least two different cell types. This complexity severely hampered in-depth studies on the regulation of Ca^{2+} transport in the distal nephron. Being aware of these limitations, we demonstrate here that the cell culture system developed, provides a useful system to unravel the intricate puzzle of regulation of transcellular Ca^{2+} transport and the possible interference with intracellular Ca^{2+} signalling.

REFERENCES

1. Van Os, C.H., Transcellular calcium transport in intestinal and renal epithelial cells, Biochim.Biophys.Acta, 906, 195, 1987.
2. Donowitz, M., Welsh, M.J., Regulation of mammalian small intestinal electrolyte secretion, in Physiology of the gastrointestinal tract, Johnson, L.R., Ed., Raven Press, New York, 1987, 1351.
3. Taylor, A., Windhager, E.E., Cytosolic calcium and its role in the regulation of transepithelial ion and water transport, in The kidney, physiology and pathophysiology, Seldin, D.W. and Giebisch, G., Eds., Raven Press, New York, 1985, 1297.
4. Bindels, R.J.M., Hartog, A., Timmermans, J., Van Os, C.H., Active Ca^{2+} transport in primary cultures of rabbit kidney collecting duct system: stimulation by 1,25-dihydroxyvitamin D_3 and PTH, Am.J.Physiol., in press, 1991.
5. Taylor, A.N., McIntosh, J.E., Bourdeau, J.E., Immunocytochemical localization of vitamin D-dependent calcium-binding protein in renal tubules of rabbit, rat and chick, Kidney Int., 21, 765, 1982.
6. Giuliano, A.R., Wood, R.J., Vitamin D-regulated calcium transport in Caco-2 cells:

unique in vitro model, Am.J.Physiol., 260, G207, 1991.

7. Bronner, F., Pansu, D., Stein, W.D., An analysis of intestinal calcium transport across the rat intestine, Am.J.Physiol., 250, G561, 1986.
8. Thomasset, M., Parkes, C.O., Cuisinier-Gleizes, P., Rat calcium-binding proteins: distribution, development and vitamin D-dependence, Am.J.Physiol., 243, E483, 1982.
9. Feher, J.L., Facilitated calcium diffusion by intestinal calcium-binding protein, Am.J.Physiol., 244, C303, 1983.
10. Costanzo, L.S., Windhager, E.E., Calcium and sodium transport by the distal convoluted tubule in the rat, Am.J.Physiol., 235, F492, 1978.
11. Shimizu, T., Yoshitomi, K., Nakamura, M., Imai, M., Effects of PTH, calcitonin and cAMP on calcium transport in rabbit distal nephron segments, Am.J.Physiol., 259, F408, 1990.
12. Rouse, D., Suki, W.N., Renal control of extracellular calcium, Kidney Int., 38, 700, 1990.
13. Yamamoto, M., Kawanobe, Y., Takashashi, H., Shimazawa, E., Kimura, S., Ogata, E., Vitamin D deficiency and renal calcium transport in the rat, J.Clin.Invest., 74, 507, 1984.
14. Gross, M., Kumar, R., Physiology and biochemistry of vitamin D-dependent calcium binding proteins, Am.J.Physiol., 259, F195, 1990.
15. Bourdeau, J.E., Lau, K., Basolateral cell membrane Ca-Na exchange in single rabbit connecting tubules, Am.J.Physiol., 258, F1497, 1990.
16. Taniguchi, S., Marchetti, J., Morel, F., Na/Ca exchangers in collecting cells of rat kidney - A single tubule Fura-2 study, Pflügers Arch., 415, 191, 1989.

PHOSPHATE TRANSPORT IN OSTEOGENIC CELLS : REGULATION AND RELATION WITH MATRIX MINERALIZATION

J. Caverzasio, T. Selz, C. Montessuit and J.-P. Bonjour

Division of Clinical Pathophysiology

University Hospital, Geneva, Switzerland

Address for correspondence :

Prof. Jean-Philippe BONJOUR
Division of Clinical Pathophysiology
Department of Medicine
University Hospital
CH - 1211 Geneva 4 (Switzerland)

Phone: (22) 22 66 20
Fax : (22) 22 73 75

TRANSPORT OF PHOSPHATE IN OSTEOBLASTIC CELLS

Osteoblasts, as competent cells for both the formation and the mineralization of the bone matrix, are directly implicated in the process of growth, modeling, and remodeling of the bony skeleton (1). Inorganic phosphate (Pi) is an essential element for the physiological functioning of osteoblasts, not only because it is an integrate component of apatite crystal but also because it can affect the production rate of the bone matrix (2). Thus, Pi deprivation interferes with normal osteoblastic function, which in turn influences the process of mineralization (3). There exists a positive relation between the plasma Pi level and the rate of skeletal growth and/or mineralization (1, 4-6). Nevertheless, several observations suggest that the availability of Pi at mineralization site level is not merely dependent upon its concentration in the systemic extracellular compartment (7-12). The translocation of Pi from the systemic to the bone extracellular compartment could be an important function of the osteoblastic cells. Therefore, it became of interest to investigate the handling and regulation of Pi by osteoblastic cells.

In a first series of experiments, the Pi transfer into the intracellular compartment was characterized in two osteoblastic cell lines, ROS 17/2.8 and UMR-106 (13, 14). It was shown to be a carrier-mediated process driven by the transmembrane electrochemical gradient of Na^+. Then it was demonstrated that Na^+-dependent Pi transport (NaPiT) in UMR-106 cells was selectively stimulated by parathyroid hormone (PTH), via a mechanism that probably involves cAMP but not the de novo synthesis of protein (14). This observation provided first experimental evidence indicating that a major osteotropic hormone could regulate the handling of Pi by osteoblastic cells. Further studies showed that IGF-1, a factor which promotes both the proliferation and differentiation of bone forming cells (15-17), can selectively stimulate NaPiT in quiescent osteoblastic cells via a mechanism that requires the de novo synthesis of protein(s) (18). The response to IGF-1, observed in the osteoblastic cell line MC 3T3 E1, was probably mediated by IGF-1 receptors. Indeed, insulin was two orders of magnitude less potent than IGF-1 to enhance NaPiT in MC 3T3 cells. Furthermore, the effect of insulin on the NaPiT system of osteoblastic cells was not selective and independent on the de novo synthesis of protein.

The foregoing knowledges prompted us to explore whether fluoride could affect the NaPiT system of osteoblastic cells. Indeed fluoride remains a unique agent that is able to increase the mass of trabecular bone (19, 20). This effect could be explained by a stimulation of the proliferation of osteoblast precursors related to an enhancement in the activity of bone cell mitogens (19, 21, 22). Interestingly, the mitogenic activity of fluoride in embyronic chick bone cells was shown to be enhanced by raising the extracellular concentration of Pi (22). These observations suggested that Pi might be a limiting factor for bone cell proliferation. Thus fluoride, through a mechanism involving the activation of bone cell mitogens, may enhance the transfer of Pi from the extra- into the intracellular compartment. The influence of fluoride on the transport of Pi was investigated in the osteoblast-like cell line UMR-106 (23). Exposure of cells to fluoride induced a dose-related stimulation of NaPiT without affecting the Na-dependent transport of alanine. The selective effect of fluoride on NaPiT was not associated with changes in cAMP, cell proliferation or alkaline phosphatase activity. However, it was completely blunted by inhibiting translational processes of protein synthesis. Furthermore, fluoride enhanced the stimulatory effect on NaPiT of various mitogens such as foetal calf serum, insulin or IGF-1. Thus, the cellular supply of Pi could be an important function controlled by growth factors and influenced by other osteogenic agents such as fluoride. Note that fluoride can enhance the activity of the Pi transporter present in the plasma membrane of osteogenic cells independently of any mitogenic action. This notion raises again the question of the physiological role of the Pi transporter present in the plasma membrane of osteoblastic cells, in the process of bone matrix formation and mineralization. Preliminary observations from our laboratory indicate that in UMR-106 cells fluoride accelerates the extra-cellular accumulation of calcium-phosphate deposited in cluster-like structures (24). Whether there is a functional relationship between the influence of fluoride on osteoblastic Pi transport and its action on mineral accumulation is an intriguing question that deserves to be further investigated.

TRANSPORT OF PHOSPHATE IN MATRIX VESICLES

Matrix vesicles (MV) derived from the plasma membrane of osteogenic cells are extracellular structures considered to play an important role in endochondral and membranous calcification (25-28). The mechanisms by which MV accumulate Pi and calcium are yet not completely understood. The presence of NaPiT in the plasma membrane of osteoblast-like cells prompted us to investigate whether isolated MV derived from chicken epiphyseal cartilage could be endowed with a similar Pi transport system (24). Pi uptake into MV was a linear process during the first minutes of incubation in presence or absence of NaCl. In presence of 150 mM NaCl, the initial rate of Pi uptake was several fold higher than with 150 mM N-methylglucamine (NMG) chloride. Other cations, such as Li, K, choline or NMG, showed partial activity to drive Pi into MV as compared to Na. In absence of Na, the Pi uptake was a non-saturable process. In presence of Na the Pi transport activity was saturable. Kinetic analysis favors a stoichiometry for the MV Na/Pi cotransport system of more than one Na ion for one Pi molecule. Interestingly, the mineralization activity of MV, as assessed by ^{45}Ca accumulation, was markedly enhanced by the presence of Na as compared to choline containing synthetic cartilage lymph (SCL) medium. Furthermore, the time course analysis of ion uptake by MV incubated in SCL medium indicated that Pi uptake might be a limiting step in the induction of the calcification process. Thus, these findings suggest that the activity of the Na-dependent Pi transport system present in MV of epiphyseal cartilage might play an important role in mineral accumulation by matrix vesicles and thus, in the induction of calcification in mineralizing tissues.

CONCLUSION

The handling of Pi by the plasma membrane of osteogenic cells appears to be operated by a Na^+-dependent saturable process which is regulated by osteotropic factors such as PTH, IGF-1 and fluoride. A Na^+-dependent Pi transport system is also present in matrix membrane vesicles of epiphyseal cartilage. In these extracellular structures the NaPiT system could represent an important mechanism involved in the process of intravesicular mineral accumulation and thus, in the initiation of calcification.

ACKNOWLEDGEMENTS

We acknowledge Mrs. Martine Buttex for her secretarial work.
The work was supported by the Swiss National Science Foundation (Grant N°. 3200.025.535) and by Chemofux Ges. (Wien, Austria).

REFERENCES

1. Kahn A.J., Fallon M.D., Teitelbaum S.L., Structure-function relationships in bone: an examination of events at cellular level. In: Peck W.A. (ed) Bone and Mineral Research Vol. 2. Elsevier Science Publishers, 1984, 125.

2. Bingham P.J., Raisz L.G., Bone growth in organ culture: effects of phosphate and other nutrients on bone and cartilage. Calcif. Tissue Res., 14,31, 1984.

3. Parfitt A.M., Mathews C., Rao D., Frame B., Kleerekoper M., Villanuery A.R., Impaired osteoblast function in metabolic bone disease. In: Frame B., Potts J.T. (eds) Clinical disorders of bone and mineral metabolism. J Excerpta Medica, Amsterdam, 1981, 116.

4. Harrison H.E., Harrison H.C., Calcium and phosphate homeostasis. In: Harrison H.E., Harrison H.C. (eds) Disorders of calcium and phosphate metabolism in childhood and adolescence. W.B. Saunders Co, Philadelphia, London, Toronto, 15, 1979.

5. Marel G.M., McKenna M.J., Frame B., Osteomalacia. In: Peck W.A. (ed) Bone and mineral research. Elsevier Science Publishers BV, 335, 1986.

6. Glorieux F.H., Marie P.J., Pettifor J.M. Delvin E.E., Bone response to phosphate salts, ergocalciferol, and calcitriol in hypophosphatemic vitamin D-resistant rickets. N. Engl. J. Med. 303, 1023, 1980.

7. De Vernejoul M.C., Marie P., Kuntz D., Gueris J., Miravet L., Rychkewaert A., Nonosteomalacic osteopathy associated with chronic hypophosphatemia. Calcif. Tissue Int. 34, 219, 1983.

8. Marie P.J., Glorieux F.H., Relation between hypomineralyzed periosteocytic lesions and bone mineralization in vitamin D-resistant rickets. Calcif. Tissue Int. 35, 443, 1983.

9. Harrison H.E., Findberg L., Rickets: primary hypophosphatemic and vitamin D-dependent varieties. J. Pediatr. 99, 84, 1981.

10. Lyles K.W., Harrolson J.M., Drezner M.K., The efficacy of vitamin D_2 and oral phosphorus therapy in X-linked hypophosphatemic rickets and osteomalacia. J. Clin. Endocrinol. Metab. 54, 307, 1982.

11. Drezner M.K., Lyles K.W., Haussler M.R., Harrelson J.M., Evaluation of a role for 1,25-dihydroxyvitamin D_3 in the pathogenesis and treatment of X-linked hypophosphatemic rickets and osteomalacia. J. Clin. Invest. 66, 1020, 1980.

12. Rasmussen H., Mazur A., Pechet M., Gertner J., Baron R., Anast C., Treatment of familial hypophosphatemic rickets. In: Norman A.W., Schaefer K., Herrath P.V., Grigoleit H.G., de Gruyter W. (eds) Vitamin D: chemical, biochemical and clinical endocrinology of calcium metabolism. Walter de Gruyter, Berlin, New York, 1982, 1219.

13. Caverzasio J., Selz T., Bonjour J.-P., Characteristics of Pi transport in osteoblast-like cells. Calcif. Tissue Int. 43, 83, 1988.

14. Selz T., Caverzasio J., Bonjour J.-P., Regulation of Na^+-dependent Pi transport by parathyroid hormone in osteoblast-like cells. Am. J. Physiol. 256, E93, 1989.

15. Schmidt C., Steiner T., Froesch E.R., Insulin-like growth factor supports differentiation of cultured osteoblast-like cells. FEBS Lett. 173, 48, 1984.

16. Hock J.M., Centrella M., Canalis E., Insulin-like growth factor I has independent effects on bone matrix formation and cell replication. Endocrinology 122, 254, 1988.

17. McCarthy T.L., Centrella M., Canalis E., Regulatory effects of insulin-like growth factors I and II on bone collagen synthesis in rat calvarial cultures. Endocrinology 124, 301, 1989.

18. Caverzasio J., Bonjour J.-P., Insulin-like growth factor 1 (IGF-1) stimulates Na^+-dependent Pi transport activity in osteoblast cells. Calcif. Tissue Int. 46, Suppl.2, A32, 1990.

19. Briançon D., Meunier P.J., Treatment of osteoporosis with fluoride, calcium and vitamin D. In : The Orthopedic Clinics of North America. Frost H.M. (ed), 629, 1981.

20. Rich C., Ensinck J., Effect of sodium fluoride on calcium metabolism of human beings. Nature 191, 184, 1961.

21. Farley J.R., Wergedal J., Baylink D.J., Fluoride directly stimulates proliferation and alkaline phosphatase activity of bone-forming cells. Science 222, 330, 1983.

22. Farley J.R., Tarbaux N., Hall S., Baylink D.J., Evidence that fluoride-stimulated ^{3}(H)-thymidine incorporation in embryonic chick calvarial cell cultures is dependent on the presence of a bone cell mitogen, sensitive to changes in the phosphate concentration, and modulated by systemic skeletal effectors. Metabolism 37, 988, 1988.

23. Caverzasio J., Selz T., Bonjour J.-P., Fluoride stimulates Pi transport in osteoblast-like cells. J. Bone Min. Res. 4, Suppl.1,, S338, 1989.

24. Caverzasio J., Very J.-M., Selz T., Bonjour J.P., Characteristics of mineral deposition (Min.D) and crystallographic analysis (CRA) of deposits in osteoblast-like (OB) cells, ROS 17/2.8, in vitro. J. Bone Min. Res. 3, suppl. 1, 1988.

25. Bonucci E., Fine structure of early cartilage calcification. J. Ultrastruct. Res. 20, 33, 1967.

26. Anderson H.C., Matrix vesicle calcification: review and update. In: W.A. Peck (ed), Bone and Min. Res., Vol. 3, Elsevier Science Publishers , 1985, 109.

27. Arsenault A.L., Hunziker E.B., Electron microscopic analysis of mineral deposits in the calcifying epiphyseal growth plate. Calcif. Tissue int. 42, 119, 1988.

28. Wuthier R.E., Mechanism of matrix vesicle-mediated mineralization of cartilage. ISI Atlas Sci. Biochem. 1, 231, 1988.

29. Montessuit C., Caverzasio J., Bonjour J.-P., Identification of a phosphate carrier in isolated matrix vesicles. Calcif. Tissue Int. 46, Suppl. 2, A24, 1990.

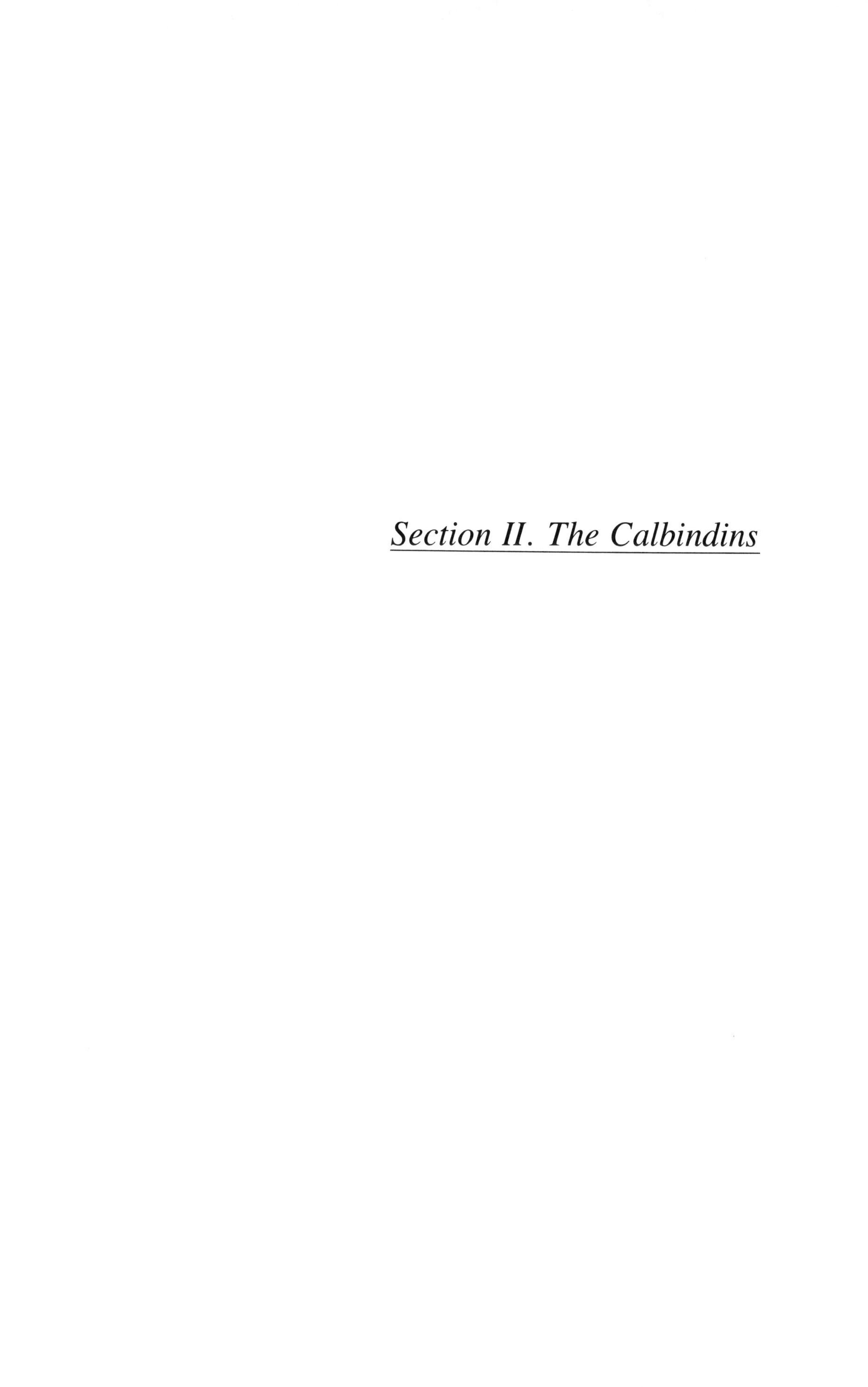

Section II. The Calbindins

ANALYSIS OF THE RAT CALBINDIN-D9K (CaBP9K) GENE : STRUCTURE, TISSUE-SPECIFIC EXPRESSION AND REGULATION

THOMASSET Monique, INSERM U120, Alliée CNRS, HOPITAL ROBERT DEBRE, 48 Bd Sérurier, 75019 PARIS, FRANCE.

I. INTRODUCTION

The calbindin-Ds (vitamin D-dependent calcium-binding proteins) are intracellular calcium-binding proteins (CaBP) which, like parvalbumin, calmodulin and troponin C, are involved in intracellular calcium homeostasis. The members of this superfamily of protein ligands thereby facilitate the action of calcium as a second messenger ([1,2]). Each calcium binding site on these proteins has the EF-hand structure described by Kretsinger ([3]).

The calbindin-Ds were discovered by Wasserman's group ([4,5]) and have calcium binding affinity constants of about 10^6 M^{-1}. The small, 9000 dalton calbindin-D9K (CaBP9K) first found in the rat small intestine, has two calcium binding sites ; the larger 30 000 dalton calbindin-D28K (CaBP28K), is most abundant in the mammalian cerebellum and kidney ([6]), and has 4 efficient calcium-binding sites, possibly 6 according to a recent paper ([7]). While mammals contain both CaBP9K and CaBP28K, birds possess only CaBP28K. CaBP9K and CaBP28K can account for as much as 2 % of the total soluble proteins in these tissues. The unique feature of CaBP9K and CaBP28K is that their synthesis is regulated by the active vitamin D metabolite, $1{,}25(OH)_2D_3$ ([6,8]), at least in the duodenum and kidney. The precise function of these two proteins remains a topic of considerable debate. Their high affinity for calcium and their high concentrations in certain tissues suggest that they may be intracellular calcium buffers, calcium carriers or modulators of enzyme activity.

We first prepared specific polyclonal antibodies to each rat CaBP ([6,9,10]), cloned the cDNA specific to each CaBP messenger RNA species ([11,12]) and isolated the CaBP9K gene ([13]). Using these tools, we have shown that each protein is encoded by a separate gene ([12,13]) ; there is no direct filiation between the two genes ([14]) and there are no similarities between the sequences of the two proteins (CaBP9K and CaBP28K) ([12,13]).

This paper will focus on the rat CaBP9K gene. Although the CaBP9K gene is mainly expressed in the duodenum ([6,8]), its product is found in the placenta and the uterus ([15,16]). CaBP9K represents 2 % of the soluble protein in the duodenum and 0.1 % in the uterus. The main difference between these two sites is that CaBP9K synthesis is under the control of $1{,}25(OH)_2D_3$ in the duodenum but not in the uterus ([17]). This article summarizes the information provided by the molecular cloning of the rat CaBP9K gene. The structure of the gene will be analyzed, together with its tissue-specific expression and its regulation by steroid hormones.

II. STRUCTURE OF THE RAT CALBINDIN-D9K (CaBP9K) GENE

Figure 1 shows the structure and organization of the CaBP9K gene. This gene is 2.5 kb long and contains 3 exons interrupted by 2 introns. The first exon contains almost the complete 5' non-coding region. The second exon codes for the calcium-binding site I. The third exon codes for calcium binding site II and the 3' untranslated region. Transcription is initiated at a single site.

```
                                            cacccacccccattttgtatt -365
tttaaaagccatgtatactatccctctgtgacgcatttttttttctgagtag -313
gaaaattattgtcaagcctaaatctcgaattccattgcttttagaaatgtgg -261
actgtgcgcaaggtcaaaattctatactgaagggaaaatatgctaaaggaga -209
agaaaagccaggttttacccttgggactctgaaccattaagcaattgttgaa -157
gtagctctctattggagcttgaccttaattgagagtcttaagcttggtctca -105
gaaaccattaatcattacccttaaatagtaaacaaagggaaatttcatatc  -53
agggtggtgtgtctgtaaagactataaaagagctcctcgtcggctcttcatc  -1
  AGACCTCACCTGTTCCTGTCTGACTCTGGCAGCACTCACTGACAGCAAG   49
 +1
CAGgtcag..........                        .........tacag   359

CAC AGA AAA ATG AGC GCT AAG AAA TCT CCC GAA GAA ATG   398
            Met Ser Ala Lys Lys Ser Pro Glu Glu Met

AAG AGC ATT TTT CAA AAA TAT GCA GCC AAA GAA GGC GAT   437
Lys Ser Ile Phe Gln Lys Tyr Ala Ala Lys Glu Gly Asp

CCA AAC CAG CTG TCC AAG GAG GAG CTG AAG CTG CTG ATT   476
Pro Asn Gln Leu Ser Lys Glu Glu Leu Lys Leu Leu Ile

CAG TCA GAG TTC CCC AGC CTC CTG AAG gtgagt .........
Gln Ser Glu Phe Pro Ser Leu Leu Lys

          .........gaagcag GCT TCA AGT ACT CTA GAC AAT  2331
                           Ala Ser Ser Thr Leu Asp Asn

CTC TTT AAA GAG CTG GAT AAG AAC GGT GAT GGA GAA GTT  2370
Leu Phe Lys Glu Leu Asp Lys Asn Gly Asp Gly Glu Val

AGC TAT GAA GAA TTC GAA GTT TTC TTC AAA AAG TTA TCA  2409
Ser Tyr Glu Glu Phe Glu Val Phe Phe Lys Lys Leu Ser

CAA TGAAGCCAGAAGAAGGAGCTCCGACACCACCTACTGATTGAATCCTAT 2460
Gln

CCAATCCCAAAGATCTAGCTGTGAGAGCAAGATACTGTTAATAAAGCAAATT 2512
CTGAGAC   atgtctctcttgtgaagtactgactcttgattacttagattt 2561
cagaggaatcttagtctgatgtgcctatttgatttgtctgaaatttcctttg 2613
ttttacattcatttatttatgtggtgtatgtattgcacatatgtacat     2661
```

Fig. 1 : Nucleotide sequence of the rat CaBP9K gene. The cap site (+1) is indicated. The 4' flanking region has been numbered negatively. GT/AG at splice junctions are double-underlined. Putative CAAT boxes and the polyadenylation signal are underlined. The TATA box is boxed.

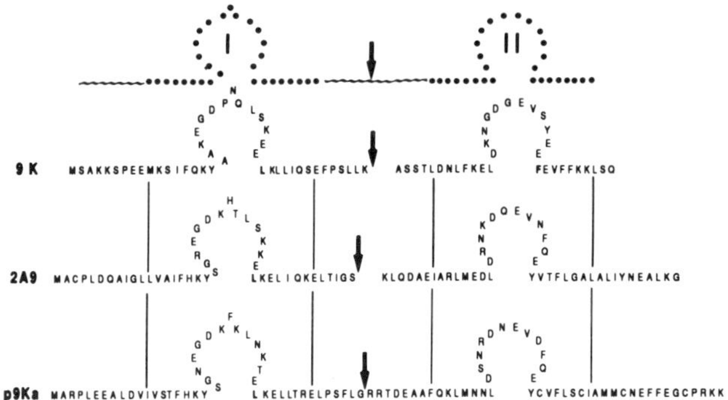

Fig. 2 : Analysis of the intron position with respect to the calcium-binding sites of the two-calcium site proteins. Each calcium-binding site (I, II) is represented by a loop, which indicates putative Ca^{++} binding sites, and by 8 aminoacid residues adjacent to the loop, which correspond to the putative regions of the helices. One intron of phase 0 ( ) is present between the 2 sites, in the interdomain region. The supposed common ancestor to these genes is shown in the upper figure.

Figure 2 outlines the CaBP9K gene. Each calcium-binding domain is encoded by a separate exon. Site I and site II are represented by a helix-loop-helix structure, with one intron between the 2 domains. Analysis of the known structures of the genes encoding other two-site calcium-binding proteins, proteins 2 A9 and p9Ka, shows similar structure (site I and site II), with one intron between each Ca-binding site. This is also true for the recently analyzed MRP8 & MRP14 genes (for review, see [13,14]). These observations support the hypothesis that these genes are derived from a common ancestor.

III. TISSUE-SPECIFIC EXPRESSION OF CALBINDIN-D9K (CaBP9K) GENE

The magnitude of CaBP9K gene expression differs from one tissue to another.

	D	Pf	Pm	U	Lg	K	Li
CaBP9K ng/mg Pr	**18250** ± 1620	**200** ± 140	**100** ± 20	**890** ± 70	**380** ± 70	**35** ± 3	**2** ± 1

Table 1 : CaBP gene expression in different tissues. D, Duodenum ; Pf, Fetal Placenta ; Pm, Maternal Placenta ; U, Uterus ; Lg, Lung ; K, kidney ; Li, Liver.

We have used Northern hybridization to show that CaBP9K is expressed as a single transcript of 0.5 kb in the rat duodenum, fetal and maternal placenta, vitelline membranes, lung and kidney. But two mRNA species appear in the uterus and the liver has no messenger.

The smaller species is similar to duodenal CaBP9K mRNA, while the larger is about 50 nucleotides longer (Figure 3). Western blot analysis shows that only a single band reacts with the CaBP9K antiserum, suggesting that only a single protein is synthesized in the uterus. This protein is identical to that of the duodenum.

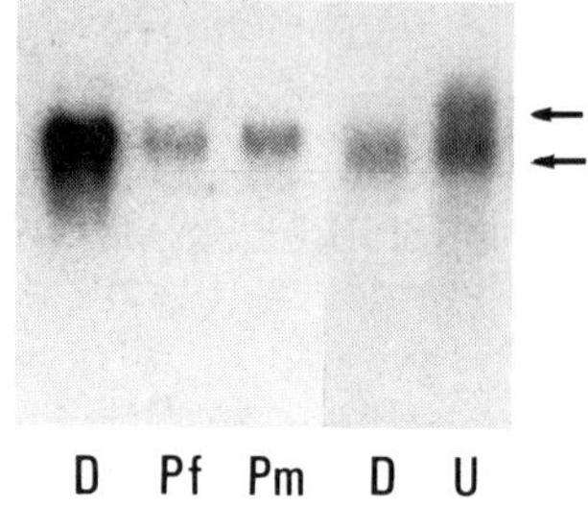

Fig. 3 : Northern blot analysis of CaBP9K mRNA D, Duodenum ; Pf, Fetal Placenta ; Pm, Maternal Placenta ; U, Uterus .

Several hypotheses, summarized in figure 4, could explain the presence of 2 mRNA species in the uterus. We have shown that there is a single transcription initiation site and a single signal for the end of transcription. Therefore the two messengers RNA are probably due to heterogeneity in the polyA tail.

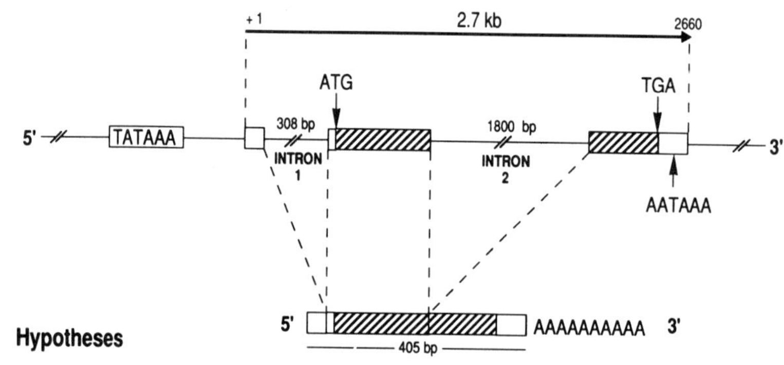

Hypotheses

- **5'** Transcription initiation : Several sites
- **3'** Internal cleavages
  End of transcription : several signals

- **Poly A tail** Different size

Fig. 4 : CaBP9K gene expression in the uterus

In order to explain the tissue-specific expression of CaBP9K gene, we mapped DNase I HS in a 28 kbp chromosomal region containing the CaBP9K gene for several rat tissues since the location of transcription regulatory elements and the positions of DNase I are correlated. Duodenal nuclei were digested with DNase I ; the resulting DNA was digested with restriction enzymes and hybridized with appropriate genomic probes.

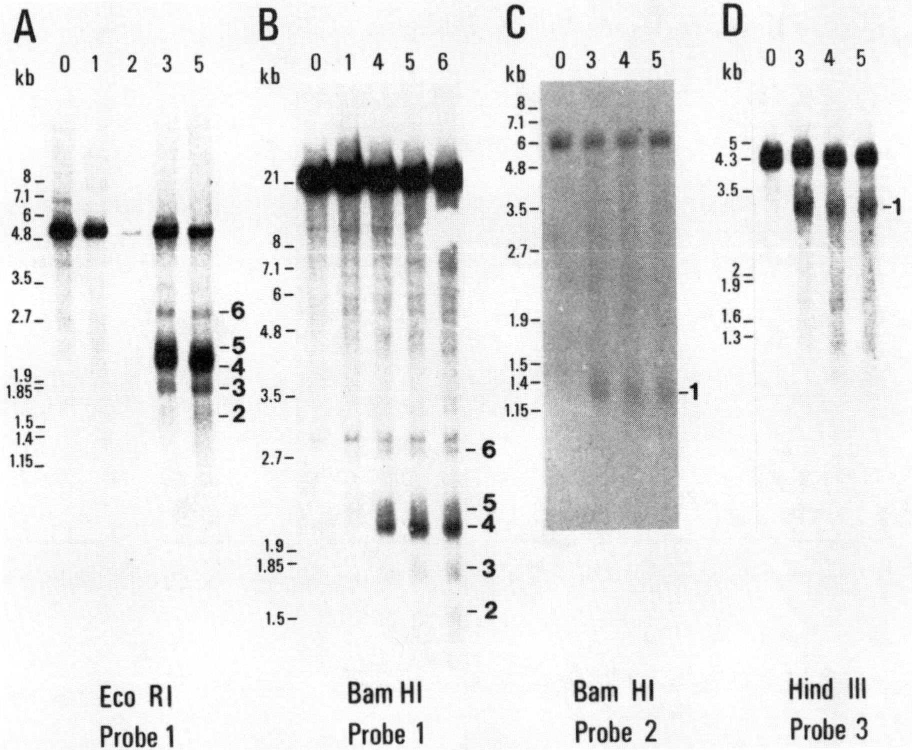

Fig. 5 : The CaBP9K hypersensitive sites in duodenal chromatin. Nuclei isolated from duodenal mucosa were digested with DNase I. Lane 0 : 0 ; Lane 2 : 40 ; Lane 3 : 80 ; Lane 4 : 120 ; Lane 5 : 160 ; Lane 6 : 320 U/ml. The DNA was extracted and digested with restriction enzymes. The restriction enzyme and probe used for each Southern blot are indicated at the bottom of each blot.

Figure 5 shows that duodenal chromatin contains 6 hypersensitive sites (HS) : 2 major ones, HS4 and HS1, and 4 minor ones, HS2, HS3, HS5 and HS6. HS4 is at the transcription site, HS1 is 3.5 kb upstream of the cap site. The minor ones, HS2, HS3, HS5 are near the promoter region and HS6 is in the second intron.

The uterine chromatin (not shown) contains no major site HS1 and major site HS4 is much less important than in duodenum chromatin. This specific distribution of DNase I hypersensitive sites is further evidence of the tissue-specific expression of the rat CaBP9K gene.

IV. TISSUE-SPECIFIC REGULATION OF THE CALBINDIN-D9K (CaBP9K) GENE

. By $1,25(OH)_2D_3$ in the rat intestine

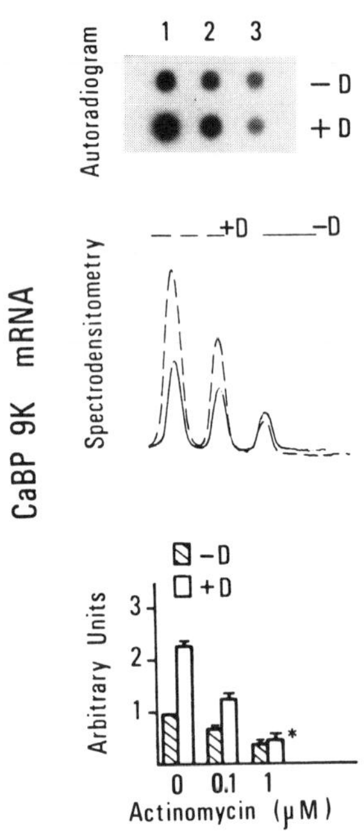

Fig. 6 : Effects of $1,25(OH)_2D_3$ and inhibitor of protein synthesis on CaBP9K mRNA. Actinomycin D was added to the culture medium in the absence of $1,25(OH)_2D_3$ (-D) or in the presence of 10^{-8}M $1,25(OH)_2D_3$ (+D). Cultures were treated for 3 h with actinomycin D (0.1 and 1 μM). Autoradiograms of 1 μg total RNA dotted and hybridized to nick-translated (^{32}P) CaBP9K. CaBP9K mRNA concentrations : analysis of autoradiograms from three separate experiments. Vertical bars represent the SEM. * $P < 0.01$ as compared to $1,25(OH)_2D_3$-treated organ culture.

In the intestine, CaBP9K gene transcription is mediated by intracellular $1,25(OH)_2D_3$ receptors that bind to the hormonal ligand and interact directly with DNA-binding sites to modulate the transcription of CaBP9K gene and other regulated genes.

$1,25(OH)_2D_3$ regulates the synthesis of CaBP9K in rats fed a vitamin D-free diet for 5 weeks from weaning (Thomasset et al 1982 ; Perret et al 1985). Under these conditions, the concentrations of CaBP9K and its mRNA in the duodenum, jejunum, ileum and caecum are greatly reduced. A single injection of $1,25(OH)_2D_3$ restores these concentrations (Thomasset et al 1982 ; Perret et al 1985). The kinetics of CaBP9K messenger accumulation of the duodenum of vitamin D-depleted rats given a single dose of $1,25(OH)_2D_3$ shows that the level begins to increase 1 hour after injection and is maximal at 6 hours, while the CaBP9K concentration increases later (Perret et al 1985).

We have developed an organ culture system for fetal rat duodena to study the regulation of CaBP9K gene by $1,25(OH)_2D_3$ in the intestine. Eighteen-day fetal rat duodena were maintained for ten days in serum-free defined medium with no exogenous $1,25(OH)_2D_3$ ([18]). Under these conditions, the cells differentiated into enterocytes, with brush borders When 10^{-8} M $1,25(OH)_2D_3$ was added to the medium the concentration of CaBP9K mRNA in the culture doubled in the first hour and increased 4-fold by 6 hours. The CaBP9K concentration increases later ([18]). Adctinomycin (1μM) inhibited the increase in CaBP9K mRNA induced by $1,25(OH)_2D_3$ at 3 hours (Figure 6).

The synthesis of CaBP9K mRNA, measured by run-on assay, increases 15 minutes after vitamin D-deficient rats are given a single $1,25(OH)_2D_3$ ([19]). This increase in CaBP9K mRNA synthesis is maximal one hour after the injection of $1,25(OH)_2D_3$. The transcription is not turned off in these vitamin D-depleted rats. However $1,25(OH)_2D_3$ has a rapid, positive effect on CaBP9K gene transcription.

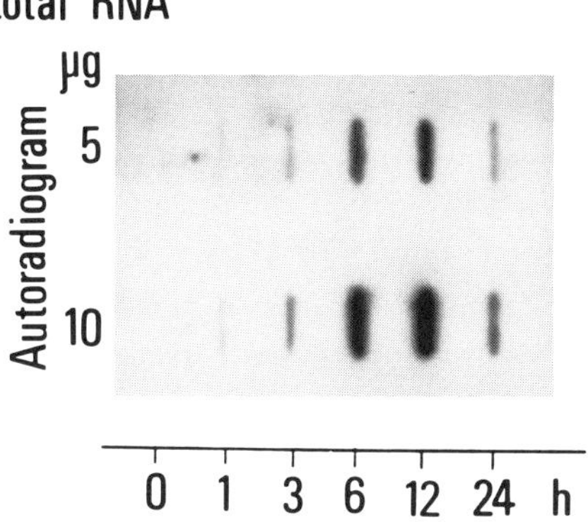

Fig. 7 : Uterine CaBP mRNA concentrations in ovariectomized rats after a single injection of 17 ß-estradiol. mRNA was measured by slot blot assay. Total RNA was analyzed 0, 1, 3, 6, 12 and 24 h after treatment. The figure shows the autoradiogram of 10 and 5 μg total RNA slotted and hybridized to the ^{32}P CaBP cDNA probe.

In the uterus, CaBP9K and its mRNA are mostly concentrated in the muscle layers and stroma in normal rats. Despite the presence of $1,25(OH)_2D_3$ receptors, exogenous $1,25(OH)_2D_3$ appears to have no effect on the uterine CaBP9K concentration ([17]). In contrast, the CaBP9K concentration is markedly reduced in ovariectomized rats and increases after injection of β-estradiol ([17]).

Slot blot analysis showed that uterine CaBP9K mRNA was undetectable in ovariectomized rats. Injecting ovariectomized rats with a single dose of estradiol greatly increased the uterine CaBP mRNA concentration (Figure 7). The messenger was detectable one hour after the treatment and significantly increased three hours after the injection of estradiol. This increase reached a maximum at 6 hours and was followed by a gradual decrease from 12 h until 96h. At this time, the concentration has returned to the basal level (not shown).

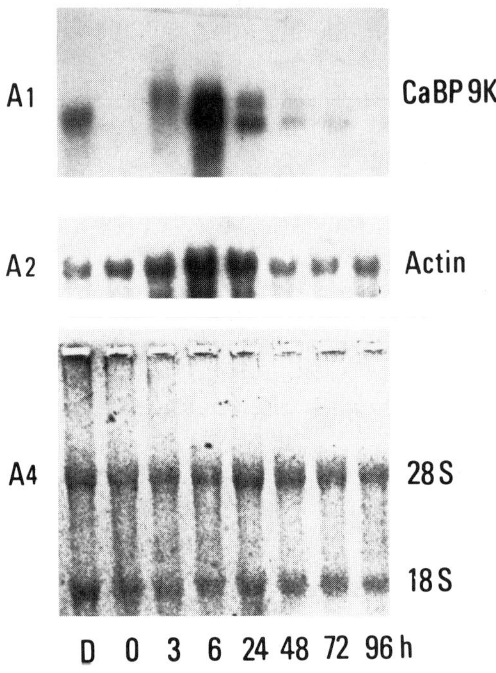

Fig. 8 : Effect of a single injection of 17 ß-estradiol on the uterine and CaBP9K mRNA concentrations in mature ovariectomized rats : Northern analysis. Autoradiograms of total RNA analyzed by Northern blot at 0, 3, 6, 24, 48, 72 and 96 h after treatment. Ten micrograms of total RNA were electrophoresed on a 2 % agarose gel, transferred to a nylon membrane, and the filter hybridized with the ^{32}P CaBP9K cDNA (A1), and the ^{32}P ß actin (A2) cDNA probes. A4 ethidium bromide control D : 10 μg total RNA extracted from the duodenum of vitamin D-deficient rat.

Since 2 messenger RNAs are present in the uterus, we have also used Northern analysis (Figure 8). The results show the existence of two estrogen-inducible mRNA species. The smaller species corresponds to the CaBP9K mRNA species found in the duodenum. The larger species is at least 50 nucleotides longer. Both mRNA species are almost undetectable in ovariectomized rats.

The 1,25$(OH)_2D_3$ receptor (VDR) belongs to the superfamily of the steroid hormone receptors. Its DNA binding domain C is very similar to that of the estrogen receptor group, including the thyroid hormone receptor (Figure 9).

1.	GRE		GGTACAnnnTGTTCT
2.	PRE		" " "
3.	ARE		" " "
4.	MRE		" " "
5.	ERE		AGGTCAnnnTGACCT
6.	EcRE		AGGGTTnnnTGCACT
7.	TRE	TC	AGGTCA---TGACCTGA
8.	RRE		" " "
9.	VDRE		GGGTGAacgGGGGCA

Fig. 9 : Consensus responsive elements for nuclear receptors, (from Beato 20). GRE : Glucocorticoids ; PRE : Progesterone ; ARE : Androgens ; MRE : Mineralocorticoids ; ERE : Estrogens ; EcRE : Ecdysone ; TRE : Thyroid Hormone ; RRE : Retinoic Acid ; VDRE : 1,25$(OH)_2D_3$.

A 1,25$(OH)_2D_3$-responsive element (VDRE) has been recently characterized by several groups within the promoter of the osteocalcin gene, the bone protein induced by 1,25$(OH)_2D_3$ (Figure 10). This sequence has some similarities with estrogen and thyroid hormone sequences.

AP - I

G T G A C T C A C C **G G G T G A** A C G **G G G G C A** T ***human***

G C C C T G C A C T **G G G T G A** A T G **A G G A C A** T ***rat***

Fig. 10 : 1,25$(OH)_2D_3$ responsive element, VDRE in osteocalcin gene, human (from Pike et al, [21]) ; rat (from Demay et al, [22]).

We have characterized the promoter region of the CaBP9K gene that contains one TATA box and several CAAT boxes (Figure 11). Analysis of 2 kb

We have characterized the promoter region of the CaBP9K gene that contains one TATA box and several CAAT boxes (Figure 11). Analysis of 2 kb upstream of the transcription initiation site showed the presence of several regulatory elements, such as HRE, which are characteristic of thyroid hormones, estrogens and other elements.

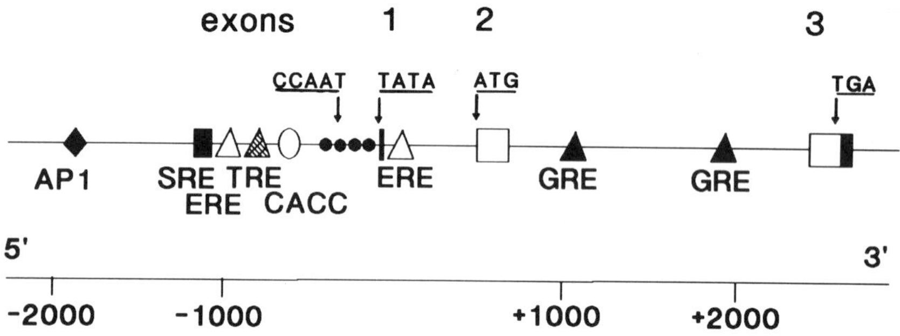

Fig. 11 : CaBP9K regulatory elements. Hormone Responsive Elements : GRE : Glucocorticoids ; ERE : Estrogens ; TRE : Thyroid Hormone ; SRE : Serum Element, AP1.

We have used cellular transfection to identify the regulatory sequences actually implicated in the control of CaBP9K gene transcription by $1,25(OH)_2D_3$.

We have checked that our CaBP9K gene construct works. The results in IEC-17 cells, derived from embryonic rat cells, show that PSVE and TkCAT, positive controls of transfection where the CAT gene is driven by a strong SV40 promoter work. CAT3 and PSB1 are negative controls where the promoter is without construct. P+CaBP is the construct with the CaBP gene promoter orientation -117 +22. P-CaBP is the construct with the antisense orientation +22-117. Only the normal sense orientation works.

This transfection assay is being used to test the biological activity of the different DNA fragments containing DNase I hypersensitive sites and hormone responsive elements.

In conclusion, the calbindin-D9K gene is a suitable model for studying the mechanisms involved in the tissue-specifc expression and hormonal regulation of a single gene by several steroid hormones.

Acknowledgments

We are grateful to C. Perret, A. Bréhier, F. L'Horset, N. Lomri and J.M. Dupret for their expert collaboration and N. Gouhier, S. Colnot and M. Lambert for their technical assistance. We thank C. Brunner and O. Parkes for their help in preparing the manuscript.

V. REFERENCES

1. Carafoli, E. , Intracellular calcium homeostasis, Ann. Rev. Biochem., 56, 395, 1987.
2. Berridge, Inositol triphosphate and diacyglycerol : two interacting second messengers, Biochem., 56, 395, 1987.
3. Kretsinger, R.H., Evolution anf function of calcium-binding proteins, Intern. Rev. Cytol., 46, 323, 1980.
4. Kallfelz, F.A., Taylor, A.N., Wasserman, R.H., Vitamin D-induced calcium factor in rat intestinal mucosa, Proc. Soc. Exp. Biol. Med., 125, 54, 1967.
5. Wasserman, R.H., Taylor, A.N., Vitamin D_3-induced calcium-binding protein in chick intestinal mucosa, Science, 152, 791, 1966.
6. Thomasset, M., Parkes, C.O., Cuisinier-Gleizes, P., Rat calcium binding protein : distribution, development and vitamin D-dependence, Am. J. Physiol., 243, E483, 1982.
7. Leathers, V.L., Linse, S., Forsén, S., Norman, A.W., Calbindin-D28K, a 1α,25-dihydroxyvitamin D3-induced calcium-binding protein, binds five or six Ca^{2+} Ions with high affinity, J. Biol. Chem., 265, 9838-9841.
8. Perret, C., Desplan, C., Thomasset, M., Cholecalcin (a 9-kDa cholecalciferol-induced calcium-binding protein) messenger RNA. Distribution and induction by calcitriol in the rat digestive tract, Eur. J. Biochem., 150, 211, 1985.
9. Intrator, S., Elion, J., Thomasset, M., Bréhier, A., Purification, immunological and biochemical characterization of rat 28 KDa cholecalcin (cholecalciferol-induced calcium-binding proteins), Biochem. J. , 231, 89, 1985.
10. Baudier, J., Glasser, N., Strid, L., Bréhier, A., Thomasset, M., Gérard, D., Purification, calcium-binding properties and conformation studies on a 28 KDa cholecalcin-like protein from bovine brain, J. Biol. Chem., 260, 10662, 1985.
11. Desplan, C., Heidmann, O., Lillile, J.W., Auffray, C., Thomasset, M., Sequence of rat intestinal vitamin D-dependent calcium-binding protein derived from a cDNA clone. Evolutionary implications,.J. Biol. Chem., 258, 13502, 1983.
12. Lomri, N., Perret, C., Gouhier, N., Thomasset, M., Cloning and analysis of calbindin-D28K cDNA and its expression in the central nervous system, Gene, 80, 87, 1989.

13. Perret, C., Lomri, N., Gouhier, N., Auffray, C., Thomasset, M., The rat vitamin D-dependent calcium-binding protein (9-KDa CaBP) gene. Complete nucleotide sequence and structural organization, Eur. J. Biochem., 172, 43-51.
14. Perret, C., Lomri, N., Thomasset, M, Evolution of the EF-hand calcium-binding protein family : evidence for exon shuffling and intron insertion. J. Mol. Evol., 27, 351, 1988.
15. Warembourg, M., Perret, C., Thomasset, M., Distribution of vitamin D-dependent calcium-binding protein messenger ribonucleic acid in rat placenta and duodenum, Endocrinology, 119, 176, 1986.
16. Warembourg, M., Perret, C., Thomasset, M., Analysis and in situ detection of cholecalcin messenger RNA (9000 Mr CaBP) in the uterus of the pregnant rat, Cell Tissue Res., 247, 5119, 1987.
17. Delorme, A.C., Danan, J.L., Acker, M.G., Ripoche, M.A., Mathieu, H., In rat uterus 17 β estradiol stimulates a calcium-binding protein similar to the duodenal vitamin D-dependent calcium-binding protein, Endocrinology, 113, 1340, 1983.
18. Bréhier, A., Thomasset, M., Stimulation of calbindin-D9K (CaBP9K) gene expression by calcium and 1,225$(OH)_2D_3$ in fetal rat duodenal organ culture, Endocrinology, 127, 580, 1990.
19. Dupret, J.M., Brun, P., Perret, C., Lomri, N., Thomasset, M., Cuisinier-Gleizes, P., Transcriptional and posttranscriptional regulation of vitamin D-dependent calcium-binding protein gene expression in the rat duodenum by 1,25-dihydroxycholecalciferol, J. Biol. Chem., 262, 16553, 1987.
20. Beato, M. Gene Regulation by steroid hormones, Cell, 56, 335, 1989.
21. Ozono, K., Liao, J., Kerner, S.A., Scott, R.A., Pike, J.W., The vitamin-D Responsive Element in the Human Osteocalcin Gene - Association with a Nuclear Protooncogene Enhancer, J. Biol. Chem., 265, 21881, 1990.
22. Morrissey, R.L., Bucci, T.J., Richard, B., Empson, N. and Lufkin, E.G. Calcium-binding protein: its cellular localization in jejunum, kidney and pancreas, Proc. Soc. Exp. Biol. Med., 149, 56, 1975.
23. Jande, S.S., Maler, L. and Lawson, D.E. Immunohistochemical mapping of vitamin D-dependent calcium-binding protein in brain, Nature, 294, 765, 1981.
24. Baimbridge, K.G. and Miller, J.J. Immunohistochemical localization of calcium-binding protein in the cerebellum, hippocampal formation and olfactory bulb of the rat, Brain. Res., 245, 223, 1982.
25. Navickis, R.J., Katzenellenbogen, B.S. and Nalbandov, A.V. Effects of the sex steroid hormones and vitamin D3 on calcium-binding proteins in the chick shell gland, Biol. Reprod., 21, 1153, 1979.
26. Bruns, M.E., Overpeck, J.G., Smith, G.C., Hirsch, G.N., Mills, S.E. and Bruns, D.E. Vitamin D-dependent clacium binding protein in rat uterus: differential effects of estrogen, tamoxifen, progesterone, and pregnancy on accumulation and cellular localization, Endocrinology, 122, 2371, 1988.
27. Taylor, A.N. Chick brain calcium binding protein: response to cholecalciferol and some developmental aspects, J. Nutr., 107, 480, 1977.
28. Hall, A.K. and Norman, A.W. Vitamin D-independent expression of chick brain calbindin-D28K, Molecular Brain Res., 9, 9, 1991.
29. Iacopino, A.M. and Christakos, S. Corticosterone regulates calbindin-D28K mRNA and protein levels in rat hippocampus, J. Biol. Chem., 265, 10177, 1990.
30. Bredderman, P.J. and Wasserman, R.H. Chemical composition, affinity for calcium, and some related properties of the vitamin D dependent calcium-binding protein, Biochemistry, 13, 1687, 1974.
31. Leathers, V.L., Linse, S., Forsen, S. and Norman, A.W. Calbindin-D28K, a 1a,25-dihydroxyvitamin D3-induced calcium-binding protien, binds five or six Ca^{2+} ions with high affinity, J. Biol. Chem., 265, 9838, 1990.

THE INTESTINAL CALBINDINS: THEIR FUNCTION, GENE EXPRESSION, AND MODULATION IN GENETIC DISEASES

R.H. Wasserman
Department and Section of Physiology
College of Veterinary Medicine
Cornell University, Ithaca, NY 14853

I. INTRODUCTION

Vitamin D is the primary factor controlling the intestinal absorption of calcium in vertebrate species [1-3]. In vitamin D deficiency, calcium absorption declines and the bone diseases, rickets in young growing animals and osteomalacia in the adult, become manifest. Aside from compromising bone metabolism, insufficient absorption of calcium over prolonged periods might affect other calcium-dependent biological processes which include muscle contraction, nerve conduction, and blood coagulation to name a few [4]. The biological significance of calcium is further accentuated by the role of Ca^{2+} as a "second messenger" in the cellular action of certain hormones and neurotransmitters [5,6].

The metabolism of vitamin D has been well studied and it is generally accepted that 1,25-dihydroxyvitamin D_3 (1,25$(OH)_2D_3$) is the most biologically active metabolite [1-3]. 1,25$(OH)_2D_3$ is synthesized in the kidney from its hydroxylated precursor, 25$(OH)D_3$. The degree of formation of 1,25$(OH)_2D_3$ reflects the calcium needs of the animal and one of the controlling factors as noted in Fig. 1 is parathyroid hormone. Also noted in Fig. 1 is the enhancing effect of 1,25$(OH)_2D_3$ on calcium transport by intestine and kidney, and the resorption of mineral from bone.

Emphasis in this chapter is given to the intestinal calcium absorptive system. 1,25$(OH)_2D_3$, like other steroid hormones, interacts with a specific receptor which, alone or in combination with other factors, induces the synthesis of proteins involved in the calcium absorptive process. In addition to the genomic action of 1,25$(OH)_2D_3$, a number of rapid effects have been recorded, including a stimulation of cyclic AMP [7,8], an alteration of the phospholipid composition [9,10] and fluidity[11,12] of the brush border membrane, and an increase in the activity of alkaline phosphatase [13-15] The rapid effect of 1,25$(OH)_2D_3$ on calcium transport by the *in vitro* double perfusion system of Nemere and Norman [16] is also considered a non-genomic response.

The exact mechanism of transport of calcium across the intestinal epithelium has not been clearly defined but three models have been proposed, as recently reviewed [17]. One model depicts calcium as entering the enterocyte, diffusing through the cytosol to be extruded by the basolateral calcium pump (CaATPase). A second model proposes that luminal calcium is

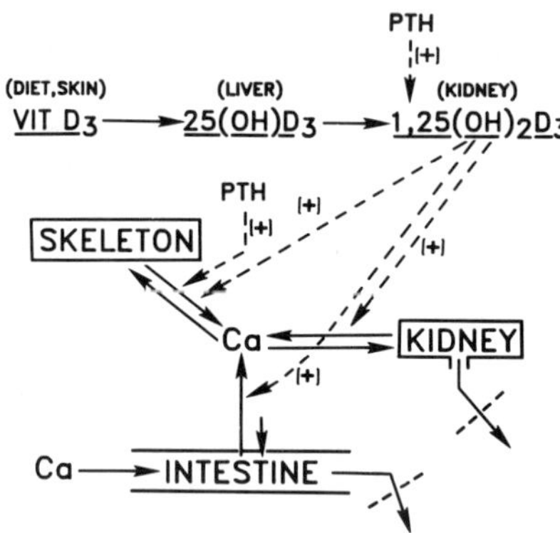

Fig.1. Vitamin D and Calcium Homeostasis. Vitamin D_3, from dietary sources or from ultraviolet light catalyzed synthesis in the skin, is converted in the liver to the 25(OH)D_3 and in the kidney to 1,25$(OH)_2D_3$. The 1,25$(OH)_2D_3$ metabolite participates in the maintenance of serum Ca^{2+} levels by increasing Ca2+ absorption from the intestine, increasing the reabsorption of Ca^{2+} from renal tubules and increasing removal of calcium from bone in concert with the parathyroid hormone (PTH). Parathyroid hormone, in addition to its bone resorptive function, stimulates the synthesis of 1,25$(OH)_2D_3$. The secretion of parathyroid hormone is elevated during periods of hypocalcemia and depressed during periods of normocalcemia and hypercalcemia. The serum Ca^{2+}-parathyroid hormone-1,25$(OH)^2D^3$ axis represents a major component of calcium homeostasis and of the feedback regulation of serum Ca^{2+} concentrations. (Not shown is the PTH effect on increasing urinary phosphate excretion). (From ref.17; reproduced with the permission of the publisher).

endocytotically acquired by the enterocytes at or near the brush border membrane, is translocated through the cell in a vesiculated form and secreted by an exocytotic process. The third proposed mechanism is the paracellular movement of calcium, the latter mechanism becoming significant at higher concentrations of intraluminal calcium. Each of these proposals has experimental and/or theoretical support.

One of the well-recognized molecular actions of 1,25$(OH)_2D_3$ on the intestine is the induction of the synthesis of the vitamin D-induced calcium binding proteins, now termed the calbindins, of which there are two classes. One class, first identified in chick intestine[18], has been named calbindin-D_{28k} [19]. The other class, first identified in rat intestine[20], has been termed, calbindin-D_{9k}. A protein closely homologous with the avian intestinal protein was early shown to be present in mammalian tissues, including the kidney [21,22] and brain [23,24].

At the outset, it should be stated that the "D" in the name, which refers to vitamin D induction of the proteins, is not strictly correct since the calbindins in the uterus of birds and mammals are apparently not vitamin D dependent but are estrogen dependent [25,26]. Brain calbindin is also not vitamin D-dependent [27,28] and considered to be constitutively expressed. Of interest, however, is the recent report showing that the protein in a specific region of the brain, the hippocampus, is decreased in adrenalectomized rats and restored by the administration of a glucocorticoid [29]. Because of these considerations, the calbindins herein will be referred to as calbindin-28k (or CaBP-28k) and calbindin-9k (or CaBP-9k), eliminating the letter 'D' for the reasons just mentioned.

II. PROPERTIES

A. BINDING AFFINITIES AND STOICHIOMETRY

Avian intestinal calbindin-28k binds 4 Ca atoms per molecule with an apparent association constant (k_a) of about 2×10^6 M^{-1} [30]. Analysis of the amino acid sequence of the protein predicted the presence of 4 calcium-binding domains (see below) and therefore coincided with the number of experimentally determined calcium-binding sites. Subsequently, Leathers et al [31] reported that avian intestinal calbindin-28k bound, not 4, but 5-6 atoms of Ca per molecule with a k_a of 6×10^7 to 2×10^8 M^{-1}, values considerably greater than those noted above. Whereas the earlier study of Bredderman and Wasserman [30] employed a buffer with an ionic strength in the physiological range (0.15 M KCl, 1 mM Pipes), Leathers et al [31] used a buffer of low non-physiological ionic strength buffer(1 mM Tris.HCl), which explains the discrepancies.

The binding affinity and the number of high affinity binding sites of the mammalian calbindin-28k have never been determined but assumed to be the same as those of the avian calbindin-28k. The basis for this assumption is the similarity between the amino acid sequences of the avian and mammalian proteins.

The mammalian calbindin-9k binds 2 Ca atoms per molecule with high affinity [32], which coincides with the number of sites predicted from its amino acid sequence(see below). The affinity of Ca binding to the high affinity sites was also determined to be about 2-4 x 10^6 M^{-1}in buffers of physiological ionic strength [32]. In low, non-physiological ionic strength buffers, the binding affinity was reported to be about 2-4 x 10^8 M^{-1} [32a].

The calbindins bind other divalent cations with varying affinities. The order of binding of the alkaline earths is Ca > Sr > Ba >> Mg [32,33]. A recent study disclosed that the calbindins binds Pb^{+2} more tightly than Ca^{+2} [34]. This could have pharmacological and toxicological implications.

B. MOLECULAR WEIGHTS AND AMINO ACID SEQUENCES

The amino acid sequence of avian CaBP-28k was determined by direct amino acid sequencing [35] and from the nucleotide sequence of cDNA-derived clones [36-39]. The protein is a single polypeptide chain consisting of 261 amino acids with a computed Mr of about 30kDa. Its N-terminal amino acid is acetylated [40]. Within the sequence are six stretches of amino acids that have sequences homologous with the E-F hand Ca-binding domains of parvalbumin, as described by Kretsinger [41]. Each E-F hand domain has an alpha-helix:Ca-binding loop:alpha-helix configuration. Of the six E-F hand domains of CaBP-28k, only 4 of the 6 are predicted to bind Ca^{2+} based on the number of oxygen-containing amino acid residues within the loop that participate in the coordinate binding of Ca2+. Two of the loops do not contain oxygen-containing amino acid residues in the required positions to bind Ca2+ or, if so, do so with low affinity.

The amino acid sequence of the mammalian calbindin-28k is homologous with the avian protein and also consists of 261 amino acids with a predicted Mr of about 30kDa [42-46]. The mammalian protein, like the avian protein, has 6 E-F hand domains and, based on the oxygen-containing residues in the binding loops, four are considered to bind Ca2+ with high affinity. Comparison of the avian and mammalian sequences show an overall identity of 79% and, when consideration is given to the relative similarities of the chemical properties of the amino acids, the homology increases to 96% [39].

The amino acid sequence of mammalian calbindins-9k was obtained by direct sequencing techniques [47-49] and deduced from the nucleotide sequence of cDNA clones [50,51]. The molecular mass of each of these mammalian proteins was determined to be about 9000 daltons and, using the E-F hand criteria of Kretsinger [41], predicted to bind 2 atoms of Ca per molecule. This prediction coincides with the experimentally determined Ca binding capacity of the high affinity binding sites of bovine, porcine, sheep and rabbit calbindins-9k [32,52,53].

The mammalian calbindin-9k proteins, although exhibiting a high degree of homology, are immunologically distinct, i.e., exhibit no antibody cross-reactivity, except between the mouse and rat and between the sheep and cow. The reason for this immunological specificity, according to Hofmann et al [48], might be due to the overall similarity among the mammalian intestinal calbindins. Antibodies raised against a mammalian calbindin in another mammalian species would only be formed against that small part of the molecule that differs between the two species. The antigenic sites apparently differ sufficiently among species to account for the general lack of cross-reactivity.

The amino acid sequence analysis of mouse intestinal calbindin-9k by tandem mass spectrometry disclosed the existence of two isoforms of the protein, differentiated by the insertion of an internal glutamine residue in one of the isoforms [54]. As suggested, this could arise from alternate splicing of a primary calbindin transcript.

The structure of the bovine CaBP-9k was determined by X-ray crystallography and the configurations of the two Ca^{2+} binding sites within the crystal structure of the protein were delineated [55,56].

III. TISSUE DISTRIBUTION

The CaBP-28k or CaBP-28k-like proteins are present in nearly all classes of tissues of mammals and birds, including epithelia, the central and peripheral nervous system, sensory organs, bones and teeth, and endocrine glands. CaBP-9k, at present, appears to have a more restricted tissue distribution, occurring only in mammals, and present in intestine, kidney, placenta, yolk sac, uterus, bones and teeth. A more comprehensive listings of the tissue distribution of the calbindins in specific cell types and different species can be found in Wasserman and Fullmer [63], Norman [64], Christakos et al [65], and Gross and Kumar [66].

IV. MOLECULAR ASPECTS OF CALBINDIN SYNTHESIS

A. THE CALBINDIN GENE

The CaBP-9k gene of rat intestine is comprised of 2.5 kilobases and has three exons separated by two introns [50]. The first calcium-binding site is located in the second exon and the second binding site is present in the third exon which also contains the 3' untranslated region. The first exon contains the entire 5' non-translated region.

The calbindin-28k gene of chick intestine spans 18.5 kilobases (kb) from the cap site to the polyadenylation site of the 2.8 kb mRNA, and is comprised of eleven coding exons with 10 intervening sequences [57]. In the upstream region of the calbindin-28k gene, a glucocorticoid-like element and a metal responsive element were identified.

The region of the chick intestinal calbindin gene which binds the receptor-1,25$(OH)_2D_3$ complex, i.e., the vitamin D responsive element (DRE), has not been identified with certainty. This also applies to other transcription factors that modulate 1,25(OH)$_2$D$_3$-mediated gene expression. By the use of the DNase I footprinting assay, Boland et al [58] noted several protected segments in the 5' upstream region, with one at -329 to -313 considered to be a candidate DRE. A prior computer analysis of the calbindin gene, using the triiodothyronine response element as the test sequence, identified a sequence in the calbindin gene in the same region that was DNase I protected in the footprinting assay [59].

With regard to the estrogen-induction of uterine CaBP-9k, Krisinger et al [60] analyzed the nucleotide sequence of the rat intestinal CaBP-9k gene and identified a 15-bp (base pair) imperfect palindrome at the start of the first intron as having a high homology with the proposed consensus sequence for the estrogen-responsive (ERE) and the glucocorticoid-responsive elements. L'Horset et al [61] verified the induction of rat uterine CaBP-9k synthesis by estrogen and, by searching the nucleotide sequence of the rat CaBP-9k gene, also identified an element similar to the ERE consensus sequence at the beginning of the first intron. Darwish et al [62] expanded on this theme and, among other approaches, demonstrated specific binding of the estrogen receptor of uterine cytosol to a nucleotide fragment containing the ERE consensus sequence of the CaBP-9k gene, using a gel retardation assay. Further a vector containing a CaBP-9K ERE-thymidine kinase-chloramphenicol acetyltransferase (CAT) gene construct was transfected into T47D cells and an estrogen-dependent expression of the CAT gene was observed.

B. GENE EXPRESSION

The synthesis of intestinal calbindin appears to have an absolute dependency on 1,25(OH)$_2$D$_3$, and the responsible cellular events are analogous to the genomic action of other steroid hormones. The 1,25(OH)$_2$D$_3$ intracellular receptor, upon binding the hormone, interacts with a chromosomal gene to induce the synthesis of the mRNA coded for calbindin or synthesis. Ribosomal translation of the calbindin mRNA yields the newly synthesized protein (Fig. 2).

"Run-on" assays have been employed to determine gene transcription as a function of time after 1,25(OH)$_2$D$_3$ dosage to vitamin D-deficient animals. In this assay, cellular nuclei are isolated and incubated in vitro under conditions which allow the synthesis of RNA to occur. The formation of CaBP RNA is assessed by use of a cDNA probe complementary to CaBP mRNA.

Theofan et al[67] examined the regulation of avian intestinal CaBP-28k gene expression by 1,25(OH)$_2$D$_3$, using the run-on assay. These investigators showed that 1,25(OH)$_2$D$_3$-mediated gene transcription occurred well within 1 hr. after 1,25(OH)$_2$D$_3$ dosage of vitamin D$_3$-deficient chicks. Peak gene transcription was at 3 hr., followed by a decline to near base levels at about 12 hr. The levels of mRNA in duodenal mucosa by Northern blot analysis were significantly increased at 3-5 hrs and maximized at 12 hrs. The first significant increase in CaBP-28k occurred at 5-6 hr. after 1,25$(OH)_2D_3$. Thus, there was a lag between the transcription of CaBP-28k gene, the formation of the mRNA, and a further lag in the synthesis of the CaBP-28k protein. These lag periods between gene transcription, the appearance of message and the translation of the message suggested post-transcriptional regulation of gene expression, which seemingly are 1,25$(OH)_2D_3$-dependent. Post-translational events, as suggested by Theofan et al [67], could include a stabilization of CaBP-28k mRNA, an effect on RNA processing and/or the efficiency of mRNA translation.Dupret et al [68] undertook similar studies on rat intestinal calbindin-9k gene expression and comparable data were reported. Gene transcription, as determined by the <u>in vitro</u> nuclei run-on assay system, preceded the bulk of the formation of the mRNA which, in turn, preceded the synthesis

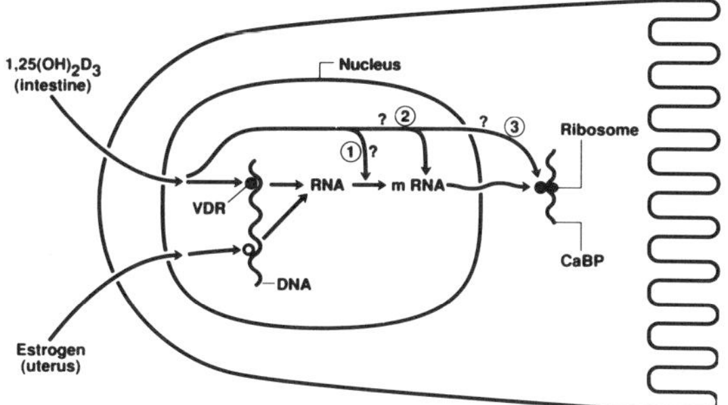

Fig.2. <u>Calbindin Gene Expression.</u> Depicted are the molecular events in calbindin gene transcription which includes the binding of 1,25$(OH)_2D_3$ to the vitamin D receptor (VDR), the interaction of the VDR-1,25$(OH)_2D_3$ with the vitamin D responsive element (VDE) in the nucleotide sequence of the calbindin gene, the transcription of the calbindin RNA, the processing of RNA to the calbindin mRNA, and translation of the message by cytosolic ribosomes. The arrows labeled (1) and (2) indicate possible post-transcriptional events which might be direct or indirect effects of 1,25$(OH)_2D_3$. Also shown as arrow (3) is the possibility of control at the translational level. Estrogen, an inducer of calbindin gene transcription in uterus, is also shown (cf. text for references).

of the CaBP-9k protein. These investigators also concluded that post-transcriptional events were involved in the CaBP-9k synthetic reaction.

Both Theofan and Norman [67] and Dupret et al [68] hypothesize the formation of a 1,25$(OH)_2D_3$-dependent protein that stabilizes CaBP mRNA. Further, this protein or another 1,25$(OH)_2D_3$-dependent protein might be involved in pre-messenger processing. The use of inhibitors of transcription and mRNA translation provided further suggestive evidence for the synthesis of a 1,25$(OH)_2D_3$-dependent CaBP-28k mRNA stabilizing factor [69,70].

The involvement of 1,25$(OH)_2D_3$-dependent post-transcriptional processing of mammalian renal CaBP-28k gene expression was also proposed by Varghese et al [71,72].

C. COTRANSCRIPTIONAL HORMONAL REGULATION OF GENE EXPRESSION

A physiological rather than a pharmacological effect of glucocorticoids on CaBP-28k synthesis and Ca transport was recently reported by Corradino and Fullmer [73]. It was shown, using an embryonic chick intestine in organ culture, that dexamethasone alone was inactive but in the presence of 1,25$(OH)_2D_3$ stimulated the synthesis of CaBP-28k mRNA and CaBP-28k above that due to 1,25$(OH)_2D_3$ alone. The uptake of Ca^{2+} by the embryonic tissue directly paralleled the changes in CaBP-28k. Actinomycin D, a transcriptional inhibitor, and cycloheximide, a translational inhibitor, depressed the formation of CaBP-28k and its mRNA by 1,25$(OH)_2D_3$ alone or in combination with dexamethasone.

Two possible mechanisms of the glucocorticoid effect were proposed. The glucocorticoid hormone-receptor complex might bind to the enhancer/promoter region of the CaBP-28k gene and thereby either augment access of the 1,25$(OH)_2D_3$-receptor complex or increase its binding affinity to the gene. The other possibility is that the glucocorticoid hormone stimulates the synthesis of the 1,25$(OH)_2D_3$ receptor.

D. RELATIONSHIP BETWEEN INTESTINAL CaBP mRNA AND CaBP

The evidence already presented suggested that regulation of calbindin synthesis might occur post-transcriptionally in RNA processing and by stabilization of translatable mRNA by an unknown factor. Supporting evidence for control at the translational level of CaBP synthesis would be indicated if there were a disparity between the relative concentrations of CaBP mRNA and the CaBP protein under different circumstances.

The view that the increase in CaBP synthesis due to dietary mineral deficiencies was at the translational level was proposed by Theofan et al [74]. It was observed that, in chicks initially fed a rachitogenic diet and then diets adequate in vitamin D but deficient in calcium or phosphorus, the intestinal concentration of CaBP-28k mRNA was unaffected but the CaBP-28k protein was elevated as compared to the normal controls. From these data, the suggestion was given that the control point in CaBP-28k synthesis was at the level of mRNA translation or the stabilization of the message but not at the gene level.

The findings of Meyer et al [75] differed from those of Theofan et al [74]. It was observed that, in chicks fed either a normal diet, a low Ca diet or a low P diet, there was a direct relationship between intestinal CaBP-28K mRNA and CaBP-28K. Fullmer [76], in a more extensive study and using the same experimental design as Theofan et al [74], also demonstrated a high correlation between the concentrations of intestinal CaBP-28k and its mRNA, as did Bar et al [77] in another series of experiments. The suggestion from these studies [75-77] is that, under the steady state conditions occurring with different dietary regimens, the intestinal concentration of CaBP is a direct function of the CaBP mRNA levels.

Chick kidney CaBP-28k responds differently to dietary mineral deficiencies than the intestine [77]. Whereas, like the intestine, a phosphorus deficiency results in the stimulation of the synthesis of renal CaBP-28k and CaBP-28k mRNA, a low Ca diet has no effect on either parameter. Likewise, Huang and Christakos [78] reported that, in rat kidney, a phosphorus deficient diet resulted in the elevation of CaBP-28k mRNA and CaBP-28k; a calcium deficient diet did not influence the levels of either CaBP-28k or its mRNA as compared to control values. Since it is known that phosphorus deficient diets increase 1,25(OH)2D3 synthesis, it seems apparent that factors other than the availability of the vitamin D hormone are involved in the control of CaBP-28k synthesis in kidney tissue, such as an increase in vitamin D receptor levels[75].

V. FUNCTION OF CALBINDIN IN EPITHELIAL TRANSPORT

A. RELATION OF CALBINDIN TO CALCIUM ABSORPTION

In severe vitamin D deficiency, CaBP-28k is virtually undetectable in chick intestine by immunological procedures and, after the administration of $1,25(OH)_2D_3$, CaBP-28k appears within the intestinal mucosa in 2-4 hrs and maximizes at 16-24 hrs. [79]. Accompanying the rise in CaBP is a parallel increase in the intestinal absorption of Ca^{2+}. In organ cultured embryonic intestine, the $1,25(OH)_2D_3$-induced synthesis of calbindin-28k occurred before or at the same time as an increase in Ca^{2+} transport was observed [80,81]. These results demonstrated a temporal relationship between CaBP induction and the vitamin D-dependent increase in Ca^{2+} absorption. It should be mentioned, however, that others[82,87] observed an increase in Ca absorption before CaBP could be detected.

A close quantitative correlation between intestinal CaBP and intestinal Ca absorption has also been shown in many other circumstances. For example, the concentration of CaBP-9k in rat intestine is highest in the duodenum, followed by the jejunum and lastly the ileum; the degree of "active" transport of Ca^{2+} follows the same pattern with a high correlation coefficient [83]. Adaptation of chicks to a low calcium or low phosphorus diet results in an increase in intestinal CaBP, with a corresponding increase in the efficiency of Ca^{2+} absorption [84]. Pharmacological doses of cortisone, known to depress Ca absorption, also depress intestinal CaBP synthesis [85]. These and other observations justifiably implicated CaBP as a key factor in the transport of Ca across the intestinal epithelium. However, other data clearly indicate that other vitamin D-dependent effects are required in addition to CaBP synthesis to account for the overall response of the intestinal calcium absorptive process to vitamin D. One example of the discordance between the CaBP content of the intestinal mucosa and Ca absorption derives from studies in which vitamin D-deficient animals are given a single dose of $1,25(OH)_2D_3$. Although the increase in intestinal CaBP and Ca absorption usually correlate with each other at early times, at later times when the absorption of Ca has declined to near base level, the mucosa tissue still retains a significant amount of CaBP [86,87]. This points to the other actions of $1,25(OH)_2D_3$ on Ca absorption which might represent limiting steps in the Ca transport path other than CaBP synthesis.

B. PROPOSED ROLES IN EPITHELIAL TRANSPORT

Essentially two non-exclusive mechanisms by which calbindin might enhance Ca absorption have been proposed: (a) CaBP might serve as a facilitator of Ca2+ diffusion from the brush border to the basolateral membrane and (b) CaBP might directly or indirectly stimulate the activity of the basolateral membrane Ca pump. These will be discussed in turn.

1. Calbindin as a diffusional facilitator

The possible functioning of CaBP as a diffusional facilitator was suggested from the theoretical analysis of the intestinal system by Kretsinger et al [88]. The essence of their analysis is, as follows. As Ca^{2+} enters the enterocyte from the lumen, a significant proportion of the entering Ca^{2+} binds to the high affinity sites of CaBP. Ca^{2+}, in the form of free Ca^{2+} and bound to CaBP, diffuses across the cell to the site of the Ca pump on the basolateral membrane. At the basolateral membrane, Ca^{2+} is actively extruded from the cell.

As the concentration of free Ca^{2+} decreases by the action of the Ca pump, it is rapidly replenished by Ca^{2+} disassociating from CaBP. In the absence of CaBP, Ca^{2+} diffuses through the cell primarily in the free form. Bronner et al [89] estimated that the free diffusion of Ca^{2+} through the cytosol of the enterocyte is too slow to account for the overall Ca^{2+} transport rate that occurs in vitamin D-replete animals and this deficit is ameliorated by the presence of CaBP. A similar estimation was made for the Ca^{2+} reabsorptive system of the renal distal convoluted tubule and a similar conclusion was offered, i.e., renal CaBP-28k allows Ca^{2+} to diffuse rapidly enough within the renal cell for the Ca pump to operate maximally [90].

The diffusional facilitator function of CaBP was tested in a multiple-compartment in vitro diffusion cell by Feher [91] and Feher et al [92]. It was shown that, at steady state, Ca^{2+} moves more rapidly through a diffusion compartment containing CaBP than through the compartment containing buffer alone. These findings support the theoretical analysis of Kretsinger et al [88] and gives credence to the calculations made by Bronner et al [89,90] on the role of CaBP in the intestinal and renal Ca2+ transport system.

Recent experiments that bear on this proposal were done in which the locality of Ca^{2+} during the course of transport across epithelial membrane was visualized [93]. The methodology made use of an ion microscope that is capable of separating different ions by mass-spectrometry and, at the same time, show where specific ions are localized in a tissue section. The stable isotope, ^{44}Ca, was employed as the transported cation to differentiate it from residual ^{40}Ca in the tissue. Experimentally, a solution containing 20 mM ^{44}Ca (as the chloride) was injected into the duodenal lumen of vitamin D-deficient or vitamin D-replete chicks. At different times, a piece of intestine was quick frozen in liquid nitrogen, the intestine cryosectioned, and then lyophilized in the frozen state. Ion microscopic analysis demonstrated a distinctly different pattern between intestinal tissue from vitamin D-deficient and vitamin D-replete chicks. In the vitamin D-deficient animals, ^{44}Ca was largely sequestered in the brush border region of the intestinal mucosa whereas, in the vitamin D-repleted chick, ^{44}Ca was distributed throughout the mucosa and present in the lamina propria. It is apparent that a vitamin D-dependent factor accounted for the difference and presumably that factor is the vitamin D-induced CaBP. In accordance with the diffusional facilitator concept of calbindin action, CaBP could very well release Ca^{2+} from the brush border region by competitive binding and increase Ca^{2+} translocation throughout the cell. Since the binding affinities of Ca2+ follow the order: brush border < calbindin < basolateral Ca pump, the directional transfer of Ca^{2+} occurs in accordance with the increasing binding affinities of components of the calcium transport path(cf.Fig.3).

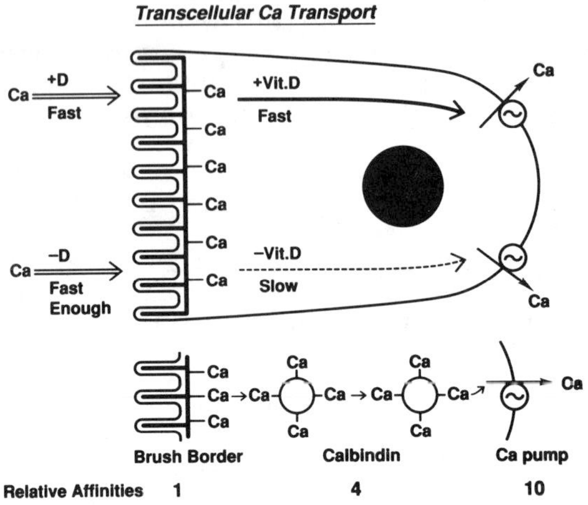

Fig.3. The transcellular diffusional-active transport model of Ca^{2+} absorption. Evidence from ion microscopic studies with a stable isotope of calcium, ^{44}Ca, suggested that Ca^{2+} entering the enterocyte is transiently sequestered by components of the brush border complex subjacent to the microvillar membrane[93]. In vit. D deficiency, Ca^{2+} diffuses slowly from this region to the basolateral membrane, where Ca^{2+} is expelled from the enterocyte by primary and secondary active transport. In the vit. D sufficient animal, the diffusional process is considerably increased. The vit. D factor proposed to account for the increased diffusion is calbindin. The relative Ca^{2+} binding affinities at the brush border complex, calbindin and the ATP-dependent Ca^{2+} pump are shown in the lower part of the figure, which indicate an up-hill gradient of binding affinities between the apical and basal-lateral poles of the enterocyte. The relative binding affinity for the brush border complex is from the data of Wilson and Lawson [94], that for calbindin-28k from Bredderman and Wasserman [30] and that for the Ca^{2+} pump. (From ref.95, reproduced with the permission of the publisher).

2. Calbindin as a possible activator of the calcium pump

In addition to CaBP acting as a diffusional facilitator, there is evidence, although somewhat controversial, that CaBP can accelerate the rate of Ca^{2+} transport across the basolateral membrane by activating the ATP-dependent basolateral Ca pump. A stimulatory effect of avian CaBP-28k on the plasma membrane Ca pump of avian intestine [95] and of the erythrocyte [96] has been reported.

However, neither Ghijsen et al [97] nor Bruns et al [98] observed a stimulatory effect of CaBP-9k on the mammalian intestinal Ca pump. In contrast, Walters [99] did demonstrate a stimulatory effect of CaBP-9k on the rat enterocyte Ca pump. This effect of CaBP-9k was, however, non-specific since parvalbumin, a Ca^{2+}-binding protein of muscle; alpha-lactalbumin, a Ca^{2+} binding protein of milk; and EGTA, a Ca^{2+} chelator, each stimulated the Ca pump. These effects might be related to an earlier observation that EGTA stimulates the erythrocyte Ca pump, and this response was termed the "EGTA effect" [100]. The explanation given for the "EGTA effect" is that the Ca^{2+} binding sites of the Ca pump are shielded by positive charges. The Ca^{2+} ion, complexed to a Ca^{2+}-binding protein or EGTA, is presumably more accessible to the Ca^{2+} binding site than Ca^{2+} alone. It is also of interest that a CaBP-9k binding domain has recently been identified in the sequence of the erythrocyte plasma membrane Ca pump [100].

Studies by the Mayo Clinic group of Borke, Kumar, Penniston and co-workers localized the plasma membrane Ca pump immunohistochemically on the basolateral membrane of the rat intestine [101], distal kidney tubules [102] and the epithelium of placenta [103]. Each of these tissues contains calbindin. The mechanism by which CaBP functions in these epithelial tissues is presumably similar to its proposed function in the intestine.

It should be re-emphasized that vitamin D, in the form of $1,25(OH)_2D_3$, does elicit other effects on the intestine in addition to inducing the synthesis of CaBP. These include an alteration of the phospholipid composition and fluidity of the brush border, the stimulated synthesis of cAMP, and the translocation of cytosolic calmodulin to the brush border complex, the calmodulin binding to a 110-kDa brush border protein (cf.[17] for references). Another recently demonstrated effect of vitamin D on the intestinal transport system is an increase the number of plasma membrane Ca pump units in the chick intestine[95] .

VI. DISEASES AND THE CALBINDINS

Certain genetic disorders and other pathological states characterized by an abnormal metabolism of calcium also affect the concentration of the intestinal calbindins. One of the earlier examples of the latter was the inhibitory effects of alloxan or streptozotocin-induced diabetes in rats on Ca absorption and a decrease in intestinal CaBP-9k[104-107]. There is a reduction in circulating levels of $1,25(OH)_2D_3$ which could account for these effects. Renal CaBP-28k was not significantly affected by streptozotocin despite the reduction in circulating levels of $1,25(OH)_2D_3$ [108]. However, in juvenile-onset diabetes in humans, renal CaBP-28k is reduced [109].

X-linked hypophosphatemic (Hyp) mice have characteristics

similar to a human form of vitamin D resistant rickets. Symptoms of this disease include a reduction in the renal tubular reabsorption of phosphate, bone disease and a reduced absorption of calcium [110]. During the postnatal period and early growth phase, the concentration of intestinal CaBP-9k is significantly lower than in normal mice. In adult Hyp mice, intestinal CaBP-9k was restored to normal levels and it was considered that the defect occurring during the maturation phase contributed to the bone disease [111]. Kidney CaBP-9k and CaBP-28k concentrations were also lower in juvenile hypophosphatemic mice as compared to normal mice, but not affected to nearly the same extent as intestinal CaBP-9k [112]. To examine the vitamin D-resistance feature of the disease, $1,25(OH)_2D_3$ was given to young Hyp and it was observed that intestinal and kidney CaBP-9k were increased but the hormone had no effect on the levels of renal CaBP-28k. A defect in the regulation of $1,25(OH)_2D_3$ production has been demonstrated in Hyp mice [113].

Osteopetrosis in man is a genetic defect characterized by an excessive deposition of bone. Seifert et al [114] studied a genetic strain of mice, the toothless (tl) mouse, characterized by osteopetrosis, absence of bone marrow cavity, and failure of tooth eruption. With respect to serum $1,25(OH)_2D_3$ concentrations and intestinal CaBP-9k, both were significantly increased in young tl mice as compared to their normal littermates. The stimulus for the excessive production of $1,25(OH)_2D_3$ was thought to be the hypophosphatemia that is present in the tl mice. Another osteopetrotic mutation in mice, the osteosclerotic (oc) mouse, was studied by the same group [115]. These mutant mice are hypophosphatemic and hypocalcemic with retarded skeletal growth and signs of rickets. There was a serum concentration of $1,25(OH)_2D_3$ six-fold higher than their normal littermates at 14-18 days of age. Intestinal CaBP-9k levels were increased considerably above that of the control group, even during the pre-weaning period (10-16 days) when, in normal mice, intestinal CaBP-9k is present only in small amounts. Renal CaBP-9k was also considerably elevated in the oc mouse. The rachitic-like state of the skeleton might be due to the hypocalcemia and hypophosphatemia that occurs despite the over-production of 1,25(OH)2D3 and excessive synthesis of CaBP-9k.

Osteopetrosis and soft tissue calcification are characteristic of another disease that affects cattle and other grazing animals in South America, Florida (USA), and in Austria and Germany(cf.ref.116 for review). In experiments designed to elucidate the nature of the calcinogenic factor, leaves of one of the plants, Solanum glaucophyllum, were fed to strontium-fed chicks. A high strontium diet inhibits $1,25(OH)_2D_3$ synthesis by renal tissue, decreases intestinal CaBP-28k to non-detectable levels and depresses Ca absorption. The dietary intake of the plant by these chicks resulted in the appearance of CaBP-28k and an increase in Ca absorption, indicative of the presence of a 1,25$(OH)_2D_3$-like substance. Subsequent biochemical analysis of the plant identified the calcinogenic factor as a glycoside of 1,25$(OH)_2D_3$. Thus, the osteopetrotic condition in these grazing animals is due to the ingestion of excessive amounts of 1,25$(OH)_2D_3$ and, unlike endogenously produced hormone, cannot be regulated.

The hypertensive rat, an inbred strain of the Wistar-Kyoto rat, the SHR strain, has been extensively studied as a model of hypertension in humans. In young SHR rats, the absorption of Ca is depressed. Cloney et al [117] noted that the concentration of intestinal CaBP-9k and alkaline phosphatase were lower in SHR rats as compared to their controls, whereas serum 1,25$(OH)_2D_3$ levels were elevated. It was concluded that, in these hypertensive rats, the defect was in the synthesis of components of the vitamin D-dependent calcium absorptive system.

VII. CONCLUSIONS

A considerable number of reports have appeared on various aspects of the functional role, tissue and species distribution, cellular localization and molecular biology of the calbindins. Only a small part of this active area of research was touched upon in this article, with emphasis given primarily to the intestinal calbindins.

Despite the progress that has already been made and the considerable amount of information that has become available, a number of interesting and significant problems remain.Still unknown is how calbindin actually operates in the intestinal calcium transport system. Does it really serve as a diffusional facilitator as has been suggested or is its role of another sort? How does calbindin function in tissues, e.g., the brain, where Ca^{2+} is not vectorially translocated across the cells of that tissue? How does 1,25$(OH)_2D_3$ activate the calbindin gene and what accessory factors, if any, are involved in gene transcription, RNA processing, and mRNA translation? How does 1,25$(OH)_2D_3$ stimulate and control, at the biochemical level, the non-genomic events that have been well-documented? These and related questions constitute challenging problems for current and future research.

VIII. REFERENCES

1. DeLuca, H.F. The vitamin D story: a collaborative effort of basic science and clinical medicine, FASEB.J.,2,224, 1988.

2. Haussler, M.R. Vitamin D receptors: nature and function, Annu.Rev.Nutr.,6,527, 1986.

3. Norman, A.W. Hormonal actions of vitamin D, Curr.Top.Cell Regul.24,35, 1984.

4. Campbell, A.K.Intracellular Calcium: Its Role as a Regulator,John Wiley & Sons,New York,1983.

5. Rasmussen, H. and Rasmussen, J.E. Calcium as intrcellular messenger: from simplicity to complexity, Current Topics in Cellular Regulation,31,1, 1990.

6. Miller, R.J. Receptor-mediated regulation of calcium channels and neurotransmitter release, FASEB J.,4,3291, 1990.

7. Corradino, R.A. Embryonic chick intestine in organ culture: interaction of adenylate cyclase system and vitamin D3-mediated calcium absorption mechanism,Endocrinology,94,1607, 1974.

8. Walling, M.W., Brasitus, T.A. and Kimberg, D.V. Elevation of cyclic AMP levels and adenylate cyclase activity in duodenal mucosa from vitamin D-deficient rats by 1 alpha,25-dihydroxycholecalciferol (1 alpha,25-(OH)2D3), Endocr.Res.Commun.,3,83, 1976.

9. Rasmussen, H., Matsumoto, T., Fontaine, O. and Goodman, D.B. Role of changes in membrane lipid structure in the action of 1, 25-dihydroxyvitamin D3, Fed.Proc.,41,72, 1982.

10. Goodman, D.B.P., Haussler, M.R. and Rasmussen, H. Vitamin D3 induced alteration of microvillar membrane lipid composition, Biochem.Biophys.Res.Comm.,46,80, 1972.

11. Brasitus, T.A., Dudeja, P.K., Eby, B. and Lau, K. Correction by 1-25-dihydroxycholecalciferol of the abnormal fluidity and lipid composition of enterocyte brush border membranes in vitamin D-deprived rats, J.Biol.Chem.,261,16404, 1986.

12. Deliconstantinos, G., Kopeikina-Tsiboukidou, L. and Tsakiris, S. Perturbations of rat intestinal brush border membranes induced by Ca2+ and vitamin D3 are detected using steady-state fluorescence polarization and alkaline phosphatase as membrane probes, Biochemical Pharmacol.,35,1633,1986.

13. Norman, A.W., Mircheff, A.K., Adams, T.H. and Spielvogel, A. Studies on the mechanism of action of calciferol III. Vitamin D-mediated increase of intestinal brush border alkaline phosphatase activity,Biochim.Biophys.Acta,215,348, 1970.

14. Nasr, L.B., Monet, J-D. and Lucas, P.A. Rapid (10-minute) stimulation of rat duodenal alkaline phosphatase activity by 1,25-dihydroxyvitamin D3, Endocrinology,123,1778, 1988.

15. Eliakim, R., Seetharam, S., Tietze, C.C. and Alpers, D.H. Differential regulation of mRNAs encoding for rat intestinal alkaline phosphatase, Am.J.Physiol.,259,G93, 1990.

16. Nemere, I. and Norman, A.W. Transcaltachia, vesicular calcium transport, and microtubule-associated calbindin-D28K: Emerging views of 1,25-dihydroxyvitamin D3-mediated intestinal calcium absorption, Mineral and Electrolyte Metabolism,16,109, 1990.

17. Wasserman, R.H. and Fullmer, C.S. On the molecular mechanisms of intestinal calcium transport, Mineral Absorption in the Monogastric GI Tract, Dintzis,F.R. and Laszlo,J.A.,Eds.,Plenum Press,1989,Chapt.5.

18. Wasserman, R.H. and Taylor, A.N. Vitamin D3-induced calcium-binding protein in chick intestinal mucosa, Science,152,791, 1966.

19. Wasserman, R.H. Nomenclature of the vitamin D-induced calcium-binding proteins, Vitamin D.A Chemical,Biochemical and Clinical Update.,Norman et al.,Eds.,Walter DeGruyter,Berlin, 1985,321.

20. Kallfelz, F.A., Taylor, A.N. and Wasserman, R.H. Vitamin D-induced calcium binding factor in rat intestinal mucosa, Proc.Soc.Exp.Biol.Med.,125,54, 1967.

21. Hermsdorf, C.L. and Bronner, F. Vitamin D-dependent calcium-binding protein from rat kidney, Biochim.Biophys.Acta,379,553, 1975.

22. Morrissey, R.L., Bucci, T.J., Richard, B., Empson, N. and Lufkin, E.G. Calcium-binding protein: its cellular localization in jejunum, kidney and pancreas, Proc.Soc.Exp.Biol.Med.,149,56, 1975.

23. Jande, S.S., Maler, L. and Lawson, D.E. Immunohistochemical mapping of vitamin D-dependent calcium-binding protein in brain, Nature,294,765, 1981.

24. Baimbridge, K.G. and Miller, J.J. Immunohistochemical localization of calcium-binding protein in the cerebellum, hippocampal formation and olfactory bulb of the rat, Brain.Res.,245,223, 1982.

25. Navickis, R.J., Katzenellenbogen, B.S. and Nalbandov, A.V. Effects of the sex steroid hormones and vitamin D3 on calcium-binding proteins in the chick shell gland, Biol.Reprod.,21,1153, 1979.

26. Bruns, M.E., Overpeck, J.G., Smith, G.C., Hirsch, G.N., Mills, S.E. and Bruns, D.E. Vitamin D-dependent calcium binding protein in rat uterus: differential effects of estrogen, tamoxifen, progesterone, and pregnancy on accumulation and cellular localization,Endocrinology,122,2371, 1988.

27. Taylor, A.N. Chick brain calcium binding protein: response to cholecalciferol and some developmental aspects, J.Nutr.,107,480, 1977.

28. Hall, A.K. and Norman, A.W. Vitamin D-independent expression of chick brain calbindin-D28K, Molecular Brain Research,9,9, 1991.

29. Iacopino, A.M. and Christakos, S. Corticosterone regulates calbindin-D28K mRNA and protein levels in rat hippocampus, J.Biol.Chem.,265,10177, 1990.

30. Bredderman, P.J. and Wasserman, R.H. Chemical composition, affinity for calcium, and some related properties of the vitamin D dependent calcium-binding protein, Biochemistry.,13,1687, 1974.

31. Leathers, V.L., Linse, S., Forsen, S. and Norman, A.W. Calbindin-D28K, a 1a,25-dihydroxyvitamin D3-induced calcium-binding protein, binds five or six Ca2+ ions with high affinity, J.Biol.Chem.,265,9838, 1990.

32. Fullmer, C.S. and Wasserman, R.H. Bovine Intestinal Calcium-Binding Protein: Cation-Binding Properties, Chemistry and Trypsin Resistance. In: Calcium Binding Proteins and Calcium Function, Wasserman, R.H. et al,Eds., Elsevier-North Holland, New York,1977, 303.

32a. Linse, S., Brodin, P., Drakenberg, T., Thulin, E., Sellers, P., Elmden, K., Grundstrom, T., and Forsen, S. Structure-function relationships in EF-hand Ca^{2+}-binding proteins. Protein engineering and biophysical studies of calbindin-D9k., Biochemistry,26,6723,1987.

33. Ingersoll, R.J. and Wasserman, R.H. Vitamin D3-induced calcium-binding protein. Binding characteristics, conformational effects, and other properties, J.Biol.Chem.,246,2808, 1971.

34. Fullmer, C.S., Edelstein, S. and Wasserman, R.H. Lead-binding properties of intestinal calcium-binding proteins, J.Biol.Chem.,260,6816, 1985.

35. Fullmer, C.S. and Wasserman, R.H. Chicken intestinal 28-kilodalton calbindin-D: complete amino acid sequence and structural considerations, Proc.Natl.Acad.Sci.USA,84,4772, 1987.

36. Hunziker, W. The 28-kDa vitamin D-dependent calcium-binding protein has a six-domain structure, Proc.Natl.Acad.Sci.USA,83,7578, 1986.

37. Hunziker, W., Siebert, P.D., King, M.W., Stucki, P., Dugaiczyk, A. and Norman, A.W. Molecular cloning of a vitamin D-dependent calcium-binding protein mRNA sequence from chick intestine, Proc.Natl.Acad.Sci.USA,80,4228, 1983.

38. Wilson, P.W., Harding, M. and Lawson, D.E. Putative amino acid sequence of chick calcium-binding protein deduced from a complementary DNA sequence, Nucleic.Acids.Res.,13,8867, 1985.

39. Gross, M.D., Kumar, R. and Hunziker, W. Expression in Escherichia coli of full-length and mutant rat brain calbindin D28. Comparison with the purified native protein, J.Biol.Chem.,263,14426, 1988.

40. Tsarbopoulos, A., Gross, M., Kumar, R. and Jardine, I. Rapid identification of calbindin-D28k cyanogen bromide peptide fragments by plasma desorption mass spectrometry, Biomedical and Environmental Mass Spectrometry,18,387, 1989.

41. Kretsinger, R.H. Calcium-binding proteins, Annu.Rev.Biochem.,45,239, 1976.

42. Takagi, T., Nojiri, M., Konishi, K., Maruyama, K. and Nonomura, Y. Amino acid sequence of vitamin D-dependent calcium-binding protein from bovine cerebellum, FEBS Lett., 201,41, 1986.

43. Yamakuni, T., Kuwano, R., Odani, S., Miki, N., Yamaguchi, K. and Takahashi, Y. Molecular cloning of cDNA to mRNA for a cerebellar spot 35 protein, J.Neurochem.,48,1590, 1987.

44. Parmentier, M., Lawson, D.E.M. and Vassart, G. Human 27-kDa calbindin complementary DNA sequence. Evolutionary and functional implications, Eur.J.Biochem.,170,207, 1987.

45. Lomri, N., Perret, C., Gouhier, N. and Thomasset, M. Cloning and analysis of calbindin-D28K cDNA and its expression in the central nervous system, Gene,80,87, 1989.

46. Wood, T.L., Kobayashi, Y., Frantz, G., Varghese, S., Christakos, S. and Tobin, A.J. Molecular cloning of mammalian 28,000 Mr vitamin D-dependent calcium binding protein (calbindin-D28K): expression of calbindin-D28K RNAs in rodent brain and kidney, DNA.,7,585, 1988.

47. Fullmer, C.S. and Wasserman, R.H. The amino acid sequence of bovine intestinal calcium-binding protein, J.Biol.Chem.,256,5669, 1981.

48. Hofmann, T., Kawakami, M., Hitchman, A.J., Harrison, J.E. and Dorrington, K.J. The amino acid sequence of porcine intestinal calcium-binding protein, Can.J.Biochem.,57,737, 1979.

49. MacManus, J.P.,Watson,D.C., and Yaguci,M.,The purification and complete amin acid sequence of the 9000-Mr Ca^{2+}-binding protein from rat placenta. Identity with the vitamin D-dependent intestinal Ca^{2+}-binding protein,Biochem.J.,235,585,1986.

50. Perret, C., Lomri, N., Gouhier, N., Auffray, C. and Thomasset, M. The rat vitamin-D-dependent calcium-binding protein (9-kDa CaBP) gene. Complete nucleotide sequence and structural organization, Eur.J.Biochem.,172,43, 1988.

51. Kumar, R., Wiegen, E. and Beecher, S.J. The molecular cloning of the complementary deoxyribonucleic acid for bovine vitamin D-dependent calcium-binding protein: Structure of the full-length protein and evidence for homologies with other calcium-binding proteins of the troponin-C superfamily of proteins, Molecular Endocrinology,3,427, 1989.

52. Bryant, D.T. and Andrews, P. Investigation of the binding of Ca2+, Mg2+, Mn2+ and K+ to the vitamin D-dependent Ca2+-binding protein from pig duodenum, Biochem.J.,219,287, 1984.

53. Chiba, K., Ohyashiki, T. and Mohri, T. Quantitative analysis of calcium binding to porcine intestinal calcium-binding protein, J.Biochem.(Tokyo.),93,487, 1983.

54. Hunt, D.F., Yates, J.R., Shabanowitz, J., Bruns, M.E. and Bruns, D.E. Amino acid sequence analysis of two mouse calbindin-D9k isoforms by tandem mass spectrometry. Protein modification by internal insertion of a single amino acid, J.Biol.Chem.,264,6580, 1989.

55. Szebenyi, D.M., Obendorf, S.K. and Moffat, K. Structure of vitamin D-dependent calcium-binding protein from bovine intestine, Nature,294,327, 1981.

56. Szebenyi, D.M. and Moffat, K. The refined structure of vitamin D-dependent calcium-binding protein from bovine intestine. Molecular details, ion binding, and implications for the structure of other calcium-binding proteins, J.Biol.Chem.,261,8761, 1986.

57. Minghetti, P.P., Cancela, L., Fujisawa, Y., Theofan, G. and Norman, A.W. Molecular structure of the chicken vitamin D-induced calbindin-D28K gene reveals eleven exons, six Ca2+-binding domains, and numerous promoter regulatory elements, Molecular Endocrinology,2,355, 1988.

58. Boland, R., Minghetti, P.P., Lowe, K.E. and Norman, A.W. Sequences near the CCAAT region and putative 1,25-dihydroxyvitamin D3-response element and further upstream novel regulatory sequences of calbindin-D28k promoter show DNase I footprinting protection, Molecular and Cellular Endocrinology,75,57, 1991.

59. Minghetti, P.P. Computer analysis of 1,25-dihydoxyvitamin D_3-receptor regulated promoters: Identification of a candidate D_3-response element, Biochem.Biophys.Res.Comm.,162,869, 1989.

60. Krisinger, J., Darwish, H., Maeda, N. and DeLuca, H.F. Structure and nucleotide sequence of the rat intestinal vitamin D-dependent calcium binding protein gene, Proc.Natl.Acad.Sci.USA,85,8988, 1988.

61. L'Horset, F., Perret, C., Brehier, A. and Thomasset, M. 17β-Estradiol stimulates the calbindin-D9k (CaBP9k) gene expression at the transcriptional and posttranscriptional levels in the rat uterus, Endocrinology, 127,2891, 1990.

62. Darwish, H., Krisinger, J., Furlow, J.D., Smith, C., Murdoch, F.E. and DeLuca, H.F. An estrogen-responsive element mediates the transcriptional regulation of calbindin D-9K gene in rat uterus, J.Biol.Chem.,266,551, 1991.

63. Wasserman, R.H. and Fullmer, C.S. Vitamin D-induced calcium-binding protein, Calcium and Cell Function II, Chapt.6,175, 1982.

64. Norman, A.W. Studies on the vitamin D endocrine system in the avian, J.Nutr.,117,797, 1987.

65. Christakos, S., Gabrielides, C. and Rhoten, W.B. Vitamin D-dependent calcium binding proteins: Chemistry, distribution, functional considerations and molecular biology, Endocrine Reviews,10,3, 1989.

66. Gross, M.D. and Kumar, R. Physiology and biochemistry of vitamin D-dependent calcium binding proteins, Am.J.Physiology, 259,F195, 1990.

67. Theofan, G., Nguyen, A.P. and Norman, A.W. Regulation of Calbindin-D28K Gene Expression by 1,25-dihydroxyvitamin D is Correlated to Receptor Occupancy, J.Biol.Chem.,261,6943, 1986.

68. Dupret, J.M., Brun, P., Perret, C., Lomri, N., Thomasset, M. and Cuisinier Gleizes, P. Transcriptional and post-transcriptional regulation of vitamin D-dependent calcium-binding protein gene expression in the rat duodenum by 1,25-dihydroxycholecalciferol, J.Biol.Chem.,262,16553, 1987.

69. Theofan, G. and Norman, A.W. Effects of alpha-amanitin and cycloheximide on 1,25-dihydroxyvitamin D3-dependent calbindin-D28K and its mRNA in vitamin D3-replete chick intestine, J.Biol.Chem.,261,7311, 1986.

70. Clemens, T.L., McGlade, S.A., Garrett, K.P., Horiuchi, N. and Hendy, G.N. Tissue-specific regulation of avian vitamin D-dependent calcium-binding protein 28-kDa mRNA by 1,25-dihydroxyvitamin D3, J.Biol.Chem.,263,13112, 1988.

71. Varghese, S., Lee, S., Huang, Y.C. and Christakos, S. Analysis of rat vitamin D-dependent calbindin-D28k gene expression, J.Biol.Chem.,263,9776, 1988.

72. Varghese, S. Transcriptional regulation and chromosomal assignment of the mammalian calbindin-D_{28k} gene, Molecular Endocrinology,3,495, 1989.

73. Corradino, R.A. and Fullmer, C.S. Positive cotranscriptional regulation of intestinal calbindin-D28k gene expression by 1,25-dihydroxyvitamin D3 and glucocorticoids, Endocrinology,128,944, 1991.

74. Theofan, G., King, M.W., Hall, A.K. and Norman, A.W. Expression of calbindin-D_{28k} mRNA as a function of altered serum calcium and phosphorus levels in vitamin D-replete chick intestine, Mol.Cell.Endocrinology, 54,135, 1987.

75. Meyer, J.,Fullmer,C.S.,Wasserman,R.H.,Komm,B.S.,and Haussler,M.R., Dietary restricion of calcium and phosphorus elicits distinctly different patterns of regulation of the mRNAs for the avian intestinal calcium binding protein and 1,25(OH)D_3 receptor, J.Bone and Mineral Research,4,S263,1989.

76. Fullmer, C.S. Regulation of intestinal calbindin-D28k gene expression: A solution hybridization study, Arch.Biochem.Biophys.,283,193, 1990.

77. Bar, A., Shami, M., Fullmer, C.S. and et al, Modulation of chick intestinal, Mol.and Cell.Endocrinology,72,23, 1990.

78. Huang, Y.C. and Christakos, S. Modulation of rat calbindin-D28 gene expression by 1,25-dihydroxyvitamin D3 and dietary alteration, Mol.Endocrinol.,2,928, 1988.

79. Wasserman, R.H., Brindak, M.E., Meyer, S.A. and Fullmer, C.S. Evidence for multiple effects of vitamin D3 on calcium absorption: response of rachitic chicks, with or without partial vitamin D3 repletion, to 1,25-dihydroxyvitamin D3, Proc.Natl.Acad.Sci.USA,79,7939, 1982.

80. Bishop, C.W., Kendrick, N.C. and DeLuca, H.F. Induction of calcium-binding protein before 1,25-dihydroxyvitamin D3 stimulation of duodenal calcium uptake, J.Biol.Chem.,258,1305, 1983.

81. Corradino, R.A. Embryonic chick intestine in organ culture: response to vitamin D 3 and its metabolites, Science, 179,402, 1973.

82. Thomasset, M., Cuisinier Gleizes, P. and Mathieu, H. 1,25-Dihydroxycholecalciferol: dynamics of the stimulation of duodenal calcium-binding protein, calcium transport and bone calcium mobilization in vitamin D and calcium-deficient rats, FEBS.Lett.,107,91, 1979.

83. Pansu, D., Bellaton, C. and Bronner, F. Developmental changes in the mechanisms of duodenal calcium transport in the rat, Am.J.Physiol.,244,20, 1983.

84. Morrissey, R.L. and Wasserman, R.H. Calcium absorption and calcium-binding protein in chicks on differing calcium and phosphorus intakes, Am.J.Physiol.,220,1509, 1971.

85. Feher, J.J. and Wasserman, R.H. Intestinal calcium-binding protein and calcium absorption in cortisol-treated chicks: effects of vitamin D3 and 1,25-dihydroxyvitamin D3, Endocrinology.,104,547, 1979.

86. Wasserman, R.H., Corradino, R.A., Feher, J. and Armbrecht, H.J. Temporal patterns of response of the intestinal absorptive system and related parameters to 1,25-dihydroxycholecalciferol, Vitamin D.Biochemical,Chemical and Clinical Aspects Related to Calcium Metabolism.,Norman et al, Eds.,Walter DeGruyter,Berlin,1977,331.

87. Spencer, R., Charman, M., Wilson, P. and Lawson, E. Vitamin D-stimulated intestinal calcium absorption may not involve calcium-binding protein directly, Nature,263,161, 1976.

88. Kretsinger, R.H., Mann, J.E. and Simmonds, J.G. Model of facilitated diffusion of calcium by the intestinal calcium binding protein,Vitamin D:Chemical,Biochemical and Clinical Endocrinology of Calcium Metabolism,Norman,A.W. et al,Eds.,Walter DeGruyter,Berlin,1982,233.

89. Bronner, F., Pansu, D. and Stein, W.D. An analysis of intestinal calcium transport across the rat intestine, Am.J.Physiol.,250,1), 1986.

90. Bronner, F. and Stein, W.D. CaBPr facilitates intracellular diffusion for Ca pumping in distal convoluted tubule, Am.J.Physiol.,255,F558, 1988.

91. Feher, J.J. Facilitated calcium diffusion by intestinal calcium-binding protein, Am.J.Physiol.,244,303, 1983.

92. Feher, J.J., Fullmer, C.S. and Fritzsch, G.K. Comparison of the enhanced steady-state diffusion of calcium by calbindin-9K and calmodulin: Possible importance in intestinal calcium absorption, Cell Calcium,10,189,1989.

93. Chandra,S.,Fullmer,C.S.,Smith,C.A.,Wasserman,R.H.,and Morrison,G.H.,Ion Microscopic imaging of calcium transport in the intestinal tissue of vitamin D-deficient and vitamin D-replete chickens: A ^{44}Ca stable isotope study,Proc.Natl.Acad.Sci.USA,87,5713,1990.

94. Wilson, P.W. and Lawson, D.E.M. Calcium binding activity by chick intestinal brush-border membrane vesicles, Pflugers Arch.,389,69, 1980.

95. Wasserman, R.H., Chandler, J.S., Meyer, S.A., Smith, C.A., Brindak, M.E., Fullmer, ,C.S., Penniston, J.T. and Kumar, R. Intestinal calcium transport and calcium extrusion processes at the basolateral membrane, J.Nutr.,in press.1991.

96. Morgan, D.W., Welton, A.F., Heick, A.E. and Christakos, S. Specific in vitro activation of Ca,Mg-ATPase by vitamin D-dependent rat renal calcium binding protein (calbindin D28K), Biochem.Biophys.Res.Commun.,138,547, 1986.

97. Ghijsen, W.E., Van Os, C.H., Heizmann, C.W. and Murer, H. Regulation of duodenal Ca2+ pump by calmodulin and vitamin D-dependent Ca2+-binding protein, Am.J.Physiol.,251,1), 1986.

98. Bruns,D.E.,Bruns,M.E., and McDonald,J.M.,ATP-dependent calcium uptake by duodenal basolateral membranes from developing rats, in Calcium-Binding Proteins in Health and Disease,Norman,A.W.,Vanaman,T.C.,and Means,A.R.,Eds.Academic Press,New York,1987,116.

99. Walters, J.R.F. Calbindin-D9k stimulates the calcium pump in rat enterocyte basolateral membranes, Am.J.Physiol.,256,G124, 1989.

100. Carafoli, E. Calcium pump of the plasma membrane, Physiological Reviews,71,129, 1991.

101. Borke, J.L., Caride, A., Verma, A.K., Penniston, J.T. and Kumar, R. Cellular and segmental distribution of Ca2+-pump epitopes in rat intestine, Pflugers Arch.,417,120, 1990.

102. Borke, J.L., Caride, A., Verma, A.K., Penniston, J.T. and Kumar, R. Plasma membrane calcium pump and 28-kDa calcium binding protein in cells of rat kidney distal tubules, Am.J.Physiol.,257,F842, 1989.

103. Borke, J.L., Caride, A., Verma, A.K., Kelley, L.K., Smith, C.H., Penniston, J.T. and Kumar, R. Calcium pump epitopes in placental trophoblast basal plasma membranes, Am.J.Physiol.,257,C341, 1989.

104. Schneider, L.E., Wilson, H.D. and Schedl, H.P. Intestinal calcium binding protein in the diabetic rat, Nature, 245,327, 1973.

105. Schneider, L.E. and Schedl, H.P. Diabetes and intestinal calcium absorption in the rat, Am.J.Physiol.,223,1319, 1972.

106. Schneider, L.E., Wilson, H.D. and Schedl, H.P. Effects of alloxan diabetes on duodenal calcium binding protein in the rat, Am.J.Physiol.,227,832, 1974.

107. Schneider, L.E., Schedl, H.P., McCain, T. and Haussler, M.R. Experimental diabetes reduces circulating 1,25-dihydroxy vitamin D in the rat, Science, 196,1452, 1977.

108. Schedl, H.P., Christakos, S., Wilson, H.D., Malkowitz, L. and Horst, R.L. Diabetes and renal calcium binding protein in the rat, Proc.Soc.Exp.Biol.Med.,177,176, 1984.

109. Orci, L., Brown, D., Roth, J. and Norman, A.W. Vitamin-D-dependent calcium binding protein content is reduced in kidneys of juvenile-onset diabetics, Lancet ,2,102, 1982.

110. Meyer, M.H., Meyer, R.A.,Jr. and Iorio, R.J. A Role for the intestine in the bone disease of juvenile X-linked hypophosphatemic mice:Malabsorption of calcium and reduced skeletal mineralization, Endocrinology,115,1464, 1984.

111. Bruns, M.E., Meyer, R.A.J. and Meyer, M.H. Low levels of intestinal vitamin D-dependent calcium-binding protein in juvenile X-linked hypophosphatemic mice, Endocrinology.,115,1459, 1984.

112. Bruns, M.E., Christakos, S., Huang, Y.C., Meyer, M.H. and Meyer, R.A. Vitamin D-dependent calcium binding proteins in the kidney and intestine of the X-linked hypophosphatemic mouse: changes with age and responses to 1,25-dihydroxycholecalciferol, Endocrinology.,121,1, 1987.

113. Tenenhouse, H.S. Abnormal renal mitochondrial 25-hydroxyvitamin D -1-alpha-hydroxylase activity in the vitamin D and calcium deficient X-linked Hyp mouse, Endocrinology ,113,816, 1983.

114. Seifert, M.F., Gray, R.W. and Bruns, M.E. Elevated 1,25-dihydroxyvitamin D3 and intestinal calbindin-D9k in the toothless rat, Am.J.Physiol.,258,E377, 1990.

115. Seifert, M.F., Gray, R.W. and Bruns, M.E. Elevated levels of vitamin D-dependent calcium-binding protein (calbindin-D9k) in the osteosclerotic (oc) mouse, Endocrinology.,122,1067, 1988.

116. Wasserman, R.H., and Nobel,T.A.,Vitamin D-related compounds as the toxic principle in calcinogenic plants, in Vitamin D: Molecular Biology and Clinical Nutrition,Norman,A.W.,Ed.,Marcell Dekker,New York,1980,455.

117. Cloney, D.L., Gray, R.W., Bruns, M.E., Burnett, S.H., Smith, M.L., Felder, R.A. and Bruns, D.E. Intestinal vitamin D-dependent calbindin-D9k and alkaline phosphatase in spontaneously hypertensive rats, Am.J.Physiol.,260,G691, 1991.

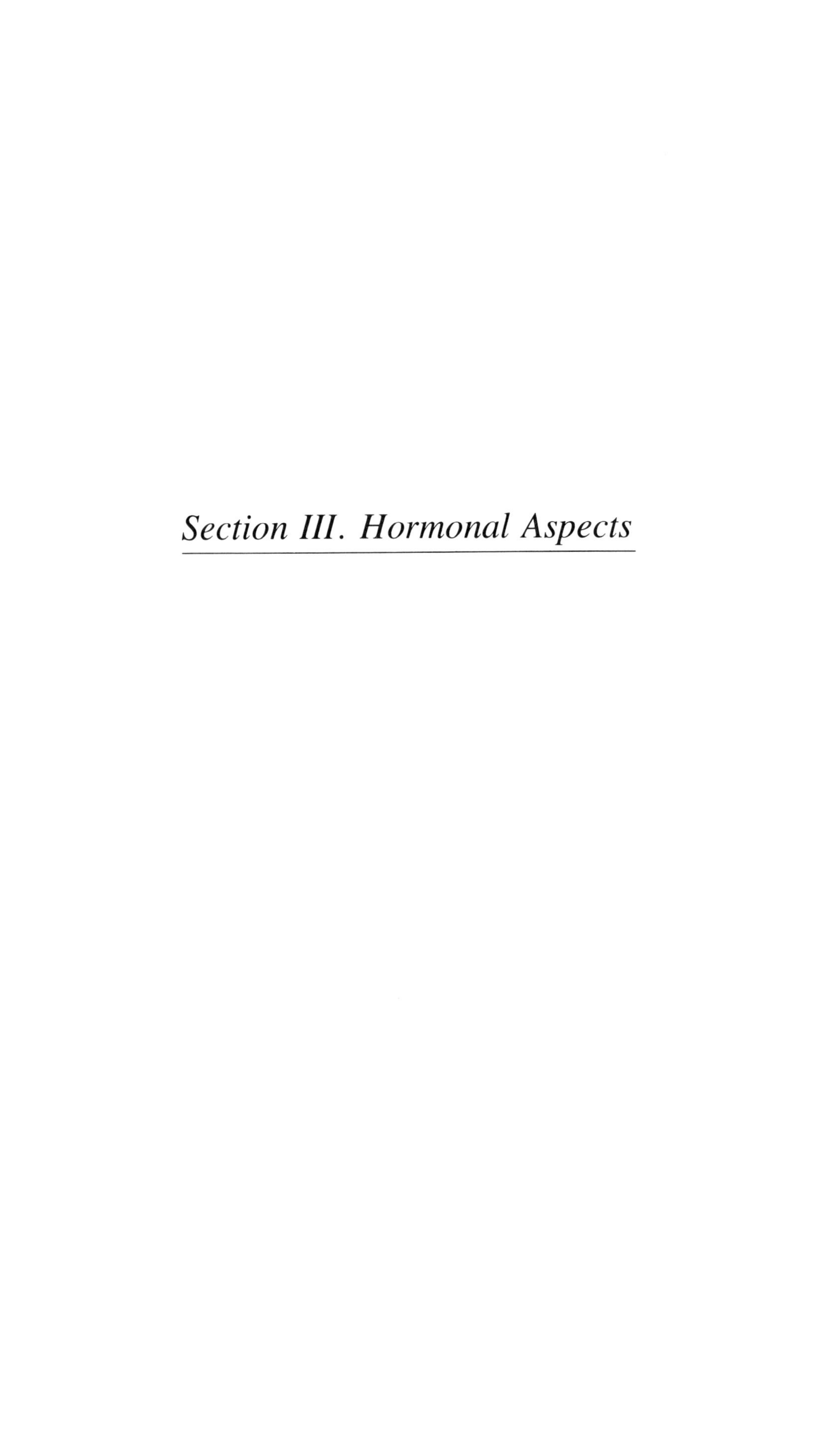

Section III. Hormonal Aspects

THE PTH/PTHrP FAMILY

T J Martin, Department of Medicine,
University of Melbourne & St Vincent's Institute
of Medical Research, St Vincent's Hospital,
Fitzroy VIC 3065, Australia

The humoral hypercalcemia of malignancy, a common syndrome in cancer, is thought to be due to the elaboration by the tumour of factor(s) which act on bone to stimulate resorption, and on kidney to decrease calcium and increase phosphorus excretion [1]. Many biochemical characteristics are shared between humoral hypercalcemia of malignancy and primary hyperparathyroidism. As clinical evidence and preliminary radioimmunoassays suggested that the responsible factor was similar to parathyroid hormone (PTH), a search to identify the PTH-like agent responsible for humoral hypercalcemia of malignancy began, concluding in 1987 with the purification, cloning, sequencing and expression of a previously unrecognized PTH-related protein (PTHrP), [2, 3], and determination of the structure of cDNA clones which represent the alternate mRNA transcripts of the PTHrP gene [4, 5].

Characterization of PTHrP

Purification of PTHrP was achieved from conditioned medium produced by a lung cancer cell line (BEN) originally derived from a patient with squamous cell carcinoma of the lung, associated with hypercalcemia [2]. The properties of this protein are similar to the activities described for tumour extracts from patients and animals with humoral hypercalcemia of malignancy [6] and in the culture medium from one such tumour cell line [5]. PTHrP resembles PTH in many of its biological actions, in that it stimulates cAMP production only in PTH target tissues and this action is prevented by antagonists of PTH. PTHrP is immunologically distinct from PTH since antisera against PTH which completely inhibit the biological effects of PTH have no effect on the activity of PTHrP [7, 8].

The limited homology at the amino-terminal region of the mature protein (eight of the first 13 identical) between PTHrP and PTH seemed sufficient to account for the similar actions of PTHrP and PTH, and indeed the amino-terminal region of PTH interacts with the PTH receptor [7]. In support of this idea, a synthetic PTHrP peptide consisting of the amino-terminal 34 residue was found to be more potent than the equivalent bovine or human PTH peptides in simulating cAMP formation in PTH-responsive osteogenic sarcoma cells [8]. Both purified and recombinant PTHrP are also more potent than either bovine or human PTH peptides in the same assay. These studies have also shown that, like PTH, PTHrP peptides of less than 30 residues from the amino-terminus have substantially less biological activity, indicating the conformational importance of the amino-terminal region [8].

Although PTH and PTHrP have similar biological activities, they differ substantially at the molecular level. For example, cDNAs encoding PTHrP have little DNA sequence homology with PTH, a surprising result in view of their amino-terminal amino acid homology. In addition, the cDNA cloning of PTHrP revealed the existence of two distinct types of 5′-untranslated sequences [3]. This, along with the detection of multiple PTHrP mRNA species, suggested the potential involvement of an alternate splicing mechanism in PTHrP gene expression which is not involved in the expression of the PTH gene.

Human PTHrP Gene

PTHrP cDNA clones have been isolated from libraries derived from a human lung cancer cell line [3] and two human renal cell carcinoma cell lines [5, 9]. They have been shown to contain essentially the same PTHrP coding sequence. However, the cDNAs differed in that they contained divergent carboxy-terminal coding regions and 5′ - and 3′ untranslated regions. This suggested that alternatively spliced mRNAs produce multiple cDNA species encoding PTHrP. The potential for alternately spliced mRNA species encoding this new hormone made it essential to isolate and characterize the human PTHrP gene in order to understand PTHrP action.

The coding region of the PTHrP gene exhibits significant structural homology to the human PTH gene, including the position of at least two introns. However, there is no significant nucleotide sequence homology to the PTH gene within the intragenic region or in the flanking genomic sequences. The PTHrP gene has been localized to human chromosome 12 [9, 10].

The biochemical similarities between PTHRP and PTH suggested that the structure of the gene encoding PTHrP could well be similar to the gene encoding human PTH [11]. The position of the exons in the 5′ end of the PTHrP gene were determined by comparing the sequences of the alternate PTHrP cDNA clones with the intron-exon boundaries of the human PTH gene. Figure 1 depicts the exon arrangement of the human PTH gene and PTHrP genes. Despite the similarity in exon structure - which is further evidence for the relatedness of the two genes - the intron-exon arrangement of the PTHrP gene is clearly more complex than that of the PTH gene.

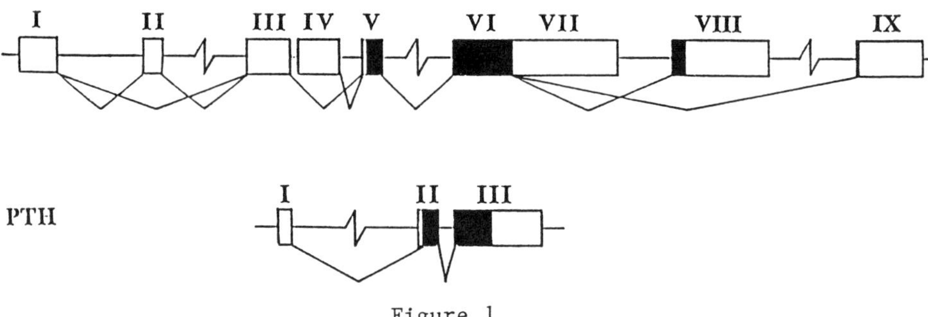

Figure 1

The PTHrP gene is composed of nine exons of which two, exons V and VI (coding the prepro molecule and the majority of the coding region, respectively), are invariant in mRNA transcripts. Exon VI comprises the majority of the coding region for the mature protein, up to residue 139, where a splice donor site is located. Read-through of exon VI into exon VII results in the introduction of a termination signal producing a protein product of 139 amino acids in length, while participation of exons VIII and IX, via alternate splicing, results in protein products of 173 and 141 amino acids, respectively (Figure 1). Thus, alternate 3′ splicing results in proteins with variable C-terminal coding regions and differing 3′ non-coding sequences. It has been suggested that 3′ untranslated sequences of some genes may play a role in development or tissue-specific gene expression. The presence of the motif ATTTA, which has been implicated in the rapid turnover of many cytokines and some protooncogene mRNA [12], in each of the 3′ untranslated sequences (i.e., those specified by exons VII, VIII and IX) is in agreement with the short half life of PTHrP mRNA. This may indicate that PTHrP has a role in either differentiation or is expressed in a tissue- specific manner, such as in the case with other mRNA species which contain this motif.

In addition to the variable 3′ exonic regions, production of multiple mRNA species by the action of alternate 5′ promoters is evident. 5′ to exons I and IV are canonical TATA consensus sequences, whilst 5′ to exon III is a GC-rich region, containing a Sp1 and two AP-2 binding sites; elements known to act as promoters in the absence of RNA II dependent polymerase. It is possible that usage of different promoters occurs in a tissue- specific manner. Consistent with this, differential use of the 5′ exons has been reported between esophageal, renal and squamous cell lines [3, 4, 5]. For example, renal cell lines utilize promoter sequences 5′ to exon I and exon IV, while squamous cell lines utilize a GC promoter region within exon III as well as the TATA sequences 5′ to exons I and IV. The extent of exon III is, at present, undefined, but it has been determined to be 240 nucleotides in length by RNase protection experiments using mRNA from a renal carcinoma cell line [5]. However by S1 and primer extension analyses this exon in esophageal cells is apparently 827 nucleotides [4], whilst in two squamous cell lines this exon is 113 nucleotides in length [13]. This apparent difference in size may result from either the tissue specific use of different promoters (e.g., AP-2, Sp1 or RNA II dependent polymerase) in these cell lines or the extension of intronic sequences present in unprocessed PTHrP mRNA transcripts.

In summary, multiple promoter sites for PTHrP in conjunction with 3′ splicing events account for the multiple mRNA species in different tissues and cancers. The use of these alternate promoters indicates complex regulation, such as has been described for the α-amylase gene, with high potential for tissue-specific processing (14).

PTHrP in Cancers

PTHrP has been detected immunohistochemically and by Northern blotting in many tumours (commonly of squamous epithelial origin) which give rise to HHM [15]. It has also been detected in parathyroid adenomata, breast tumours from normocalcemic patients [16], hypertrophic breast tissue and in a percentage of hypercalcemic patients with HTLV 1 lymphomas. In normal subjects PTHrP can be identified in high turnover epithelial tissues in skin and lactating breast (and breast milk) [15, 17]. It is also present in many early fetal tissues including fetal epithelia, muscle, bone, kidney, parathyroids and placenta, and its expression by embryonal carcinoma cells suggests a possible role in developmental processes. A role for PTHrP in placental calcium transport has been demonstrated, located beyond the first 34 amino-acids but within PTHrP (1-84) [18, 19]. Such an action may also have relevance to calcium transport into milk by breast epithelium, and thus also in the local control of the calcium environment critical for epithelial growth and differentiation. This hypothesis is supported by localisation studies which indicate its presence in highest concentrations in areas of active growth in normal and cancerous epithelial cells and in normal skin which undergoes rapid turnover.

Radioimmunoassays are currently not sensitive enough to identify PTHrP convincingly in the plasma of normal subjects, but elevated levels of PTHrP are detected in a high proportion of patients with hypercalcemia in cancer [20, 21]. Furthermore, antibodies to N-terminal regions of PTHrP can reduce the hypercalcemia and prevent the bone abnormalities in a nude mouse model of HHM [22, 23]. There are however, some discrepancies between the features of HHM and hyperparathyroidism [1] which may relate either to interactions with other tumour factors which may be co-secreted with PTHrP, or possibly to actions mediated via regions of the PTHrP molecule beyond the first 34 amino-acids. For example we have noted altered renal handling of bicarbonate by the rat kidney perfused with PTHrP (1-141) as compared to PTHrP (1-34) [24].

Role of PTHrP in Fetal Development

The regulation of fetal calcium metabolism is of considerable interest, but little is yet known of the mechanisms controlling delivery of calcium to the fetus for skeletal development. Parathyroid hormone was suspected to be involved, but there is much conflicting data on PTH levels in maternal and fetal circulations. The data has suffered from the shortcomings of the radioimmunoassays for PTH. Comparisons of bioactivity and immunoreactivity have been more revealing, in that assays of cord blood and fetal blood from sheep have shown that PTH-like bioactivity was greater than PTH immunoreactivity. In addition, fetal and cord blood PTH bioactivity exceeded maternal PTH bioactivity which was slightly elevated over non-pregnant adults [25]. In the latter study it was shown that in the first 24 to 48 hours of life immunoreactive and bioactive PTH increased as calcium decreased in both premature and term infants, both activities returning to normal in line with calcium levels. The discrepancy of immunoreactive and bioactive PTH in fetal and cord blood however suggested that PTHrP may be produced by the fetus and/or the placenta and be the important regulator of fetal calcium levels.

It has long been known that fetal calcium levels are raised relative to maternal calcium indicating the active transport of calcium across the placenta from mother to fetus. Care et al [26] studying placental calcium transport in sheep, established that fetal parathyroidectomy abolished the calcium gradient. However PTH could not restore the gradient, suggesting that a fetal parathyroid factor other than PTH could be involved. Furthermore neither calcitonin nor 1,25 $(OH)_2$ vitamin D3 could elicit any response in this model. It has now been demonstrated that early placenta from sheep, [18] and also fetal sheep parathyroids contain PTHrP-like material as detected by bioassay, Western blotting, radioimmunoassay and immunocytochemistry [19]. Purified PTHrP and recombinant PTHrP, but not PTHrP (1-34) can restore the calcium gradient in the perfused placenta from the parathyroidectomized sheep fetus, as can extracts of fetal parathyroids and placenta from early in gestation [18]. The current hypothesis is that PTHrP is produced early in gestation by the placenta and later by the fetal parathyroids. The mechanisms of PTHrP stimulation and regulation of placental calcium transport remain to be elucidated, but the effect of PTHrP on the placenta is maintained by regions of PTHrP molecule not homologous with PTH and presumably, therefore, via a unique receptor.

PTHrP has also been detected in maternal milk at very high levels (100 ng/ml), [17, 27] being found in milk from a cross-section of vertebrate species including monotremes and marsupials [17]. The latter has a very immature skeleton when it is born. Suckling and prolactin have been shown to stimulate PTHrP mRNA in mammary tissue [28], but it is not yet clear whether the presence of PTHrP is related to milk production or whether it is necessary for the neonate. There is evidence that extracellular calcium levels are important in mammary cell growth and differentiation [29]. More recently it has been reported that mRNA for PTHrP in the rat uterus is elevated prior to parturition but as yet the functional relevance of this observation is unknown [30].

It is fascinating that fetal calcium regulation is controlled by a molecule with certain PTH activities, but also apparently though extra actions associated with other regions of the molecule, which also raise extracellular calcium levels. The physiological function of PTH is to raise the plasma calcium by the means available to it, i.e. by promoting bone resorption, reducing calcium excretion and indirectly promoting intestinal calcium absorption (via 1, 25-dihydroxyvitamin D_3). In the fetus the predominant means available to keep the extracellular fluid calcium elevated are placental calcium transport and restriction of calcium excretion by the fetal kidney, both of which are brought about by PTHrP. Thus PTHrP may function in the fetus as the hormone of the fetal parathyroid gland carrying out the function which PTH does post-natally. The major function of PTHrP may well be lost or diminished in adult mammals, having relevance only in certain cancers and possibly in some parathyroid diseases. The question arises whether PTHrP is the evolutionary predecessor of PTH and might therefore be expressed in lower vertebrates where it acts as a regulator of extracellular calcium.

References

1. Martin, T. J. and Atkins, D. Biochemical regulators of bone resorption and their significance in cancer. Essays Med. Biochem., 4, 49, 1979.

2. Moseley, J. M., Kubota, M., Diefenbach-Jagger, H., Wettenhall, R. E. H., Kemp, B. E., Suva, L. J., Rodda, C. P., Ebeling, P. R., Hudson, P. J., Zajac, J. D., Martin, T. J. Parathyroid hormone-related protein purified from a human lung cancer cell line. Proc. Natl. Acad. Sci. USA, 84, 5048, 1987.

3. Suva, L. J., Winslow, G. A., Wettenhall, R. E. H., Kemp, B. E., Hudson, P. J., Diefenbach-Jagger, H., Moseley, J. M., Rodda, C. P., Martin, T. J., Wood, W. I. A parathyroid hormone-related protein implicated in malignant hypercalcemia: cloning and expression. Science, 237, 893, 1987.

4. Mangin, M., Ikeda, K., Dreyer, B. E., Milstone, L., Broadus, A. E. Two distinct tumor-derived parathyroid hormone-like peptides result from alternative ribonucleic acid processing. Mol. Endocrinol., 2, 1049, 1988.

5. Thiede, M. A., Strewler, R. A., Nissenson, R. A., Rosenblatt, M., Rodan, G. A. Human renal carcinoma expresses two messages encoding a parathyroid hormone like peptide: evidence for the alternate splicing of a single copy gene. Proc. Natl. Acad. Sci, USA, 85, 4605, 1988.

6. Stewart, A. F., Insogna, K. L., Goltzman, D., Broadus, A. E. Identification of adenylate cyclase-stimulating activity and cytochemical glucose-6-phosphate dehydrogenase stimulating activity in extracts of tumors from patients with humoral hypercalcemia of malignancy. Proc. Natl. Acad. Sci. USA, 80, 1454, 1983.

7. Martin, T. J., Allan, E. H., Caple, I. W., Care, A. D., Danks, J. A., Diefenbach-Jagger, H., Ebeling, P. R., Gillespie, M. T., Hammonds, G., Heath, J. A., Hudson, P. J., Kemp, B. E., Kukreja, S. C., Moseley, J. M., Ng, K. W., Raisz, L. G., Rodda, C. P., Simmons, H. A., Suva, L. J., Wettenhall, R. E. H. and Wood, W. I. Parathyroid hormone-related protein: isolation, molecular cloning and mechanism of action. Recent Progress in Hormone Research, 45, 467, 1989.

8. Kemp, B. E., Moseley, J. M., Rodda, C. P., Ebeling, P. R., Wettenhall, R. E. H., Stapleton, D., Diefenbach-Jagger, H., Ure, F., Michelangeli, V. P., Simmons, H. A., Raisz, L. G. and Martin, T. J. Parathyroid hormone-related protein of malignancy: active synthetic fragments. Science (Washington DC), 238, 1568, 1987.

9. Mangin, M., Webb, A. C., Dreyer, B. E., Posillico, J. T., Ikeda, K., Weir, E. C., Stewart, A. F., Bander, N. H., Milstone, L., Barton, D. E., Francke, U. and Broadus, A. E. Identification of a cDNA encoding a parathyroid hormone-like peptide from a human tumor associated with humoral hypercalcemia of malignancy. Proc. Natl. Acad. Sci. USA, 85, 597, 1988.

10. Suva, L.J., Mather, K.A., Gillespie, M.T., Webb, G.C., Ng, K.W., Winslow, G.A., Wood, W.I., Martin, T.J., Hudson, P.J. Structure of the 5′ flanking region of the gene encoding human parathyroid-hormone-related protein (PTHrP). Gene, 77, 95, 1989.

11. Vasicek, T.J., McDevitt, B.E., Freeman, M.W., Fennick, B.J., Hendy, G.N., Potts, J.T. Jr., Rich, A., Kronenberg, H.M. Nucleotide sequence of the human parathyroid hormone gene. Proc. Natl. Acad. Sci. USA, 80, 2127, 1983.

12. Shaw, G., Karmen, R. A conserved AU sequence from the 3′ untranslated region of GM-CSF mRNA mediates mRNA degradation. Cell, 46, 659, 1986.

13. Yasuda, T., Banville, D., Hendy, G.N., Goltzman, D. Characterization of the human parathyroid hormone-like peptide gene. J. Biol. Chem., 264, 7720, 1989.

14. Young, R.A., Hagenbuchle, O., Schibler, V. A single mouse α-amylase gene specifies two different tissue-specific mRNAs. Cell, 23, 451, 1981.

15. Danks, J.A., Ebeling, P.R., Hayman, J., Chou, S.T., Moseley, J.M., Dunlop, J., Kemp, B.E. and Martin, T.J. Parathyroid hormone-related protein: immunohistochemical localization in cancers and in normal skin. J. Bone Mineral Res., 4, 273, 1989.

16. Southby, J., Kissin, M.W., Danks, J.A., Hayman, J.A., Moseley, J.M., Henderson, M.A., Bennett, R.C. and Martin, T.J. Immunohistochemical localization of parathyroid hormone-related protein in human breast cancer. Cancer Res. 50, 7710, 1990.

17. Law, F.M.K., Moate, P.J., Leaver, D.D., Diefenbach-Jagger, H., Grill, V., Ho, P.W.M., Martin, T.J. Parathyroid hormone-related protein in milk and its correlation with bovine milk calcium. J. Endocrinol., 128, 21, 1991.

18. Rodda, C.P., Kubota, M., Heath, J.A., Ebeling, P.R., Moseley, J.M., Care, A.D., Caple, I.W. and Martin, T.J. Evidence for a novel parathyroid hormone-related protein in fetal lamb parathyroid glands and sheep placenta: comparisons with a similar protein implicated in humoral hypercalcemia of malignancy. J. Endocrinol., 117, 261, 1988.

19. Abbas, S.K., Pickard, D.W., Illingworth, D., Storer, J., Purdie, D.W., Moniz, C., Dixit, M., Caple, I.W., Ebeling, P.R., Rodda, C.P., Martin, T.J. and Care, A.D. Measurement of parathyroid hormone-related protein in extracts of fetal parathyroid glands and placental membranes. J. Endocrinol. 124, 139, 1990.

20. Burtis, W.J., Brady, T.G., Orloff, J.J., Ersbak, J.B., Warrell, R.P., Olson, B.R., Wu, T.L., Mitnick, M.E., Broadus, A.E. and Stewart, A.F. Immunochemical characterization of circulating parathyroid hormone-related protein in patients with humoral hypercalcemia of cancer. New Engl. J. Med. 322, 1106, 1990.

21. Grill, V., Ho, P., Body, J.J., Donohoo, G., Ferber, I., Johanson, N., Orf, J.W., Lee, S.C., Kukreja, S.C., Moseley, J.M. and Martin, T.J. Parathyroid hormone-related protein: elevated levels both in humoral hypercalcemia of malignancy and in hypercalcemia complicating metastatic breast cancer. (Submitted).

22. Kukreja, S.C., Shevrin, D.H., Wimbiscus, S.A., Ebeling, P.R., Danks, J.A., Rodda, C.P., Wood, W.I. and Martin, T.J. Antibodies to parathyroid hormone-related protein lower serum calcium in athymic mouse models of malignancy-associated hypercalcemia due to human tumors. J. Clin. Invest., 82, 1798, 1988.

23. Kukreja, S.C., Rosol, T.J., Wimbiscus, S.A., Shevrin, D., Grill, V., Barengolts, E.I. and Martin, T.J. Tumor resection and antibodies to parathyroid hormone-related protein cause similar changes on bone histomorphometry in hypercalcemia of cancer. Endocrinology, 127, 305, 1990.

24. Ellis, A.G., Adam, W.R. and Martin, T.J. Comparison of the effects of parathyroid hormone (PTH) and recombinant PTH-related protein on bicarbonate excretion by the isolated perfused rat kidney. J. Endocrinol. 126, 403, 1990.

25. Allgrove, J., Adam, S., Manning, R.M. and O'Riordan, J.L.H. Cytochemical bioassay of parathyroid hormone in maternal and cord blood. Arch. Dis. Childhood, 60, 110, 1985.

26. Care, A.D., Caple, I.W. and Pickard, D.W. The roles of the parathyroid and thyroid glands on calcium homeostasis in the ovine fetus, in The Physiological Development of the Fetus and Newborn, Jones, C.T., Nathaniels, P., Eds., Academic Press, New York, p. 135, 1985.

27. Budayr, A.A., Halloran, B.P., King, J.C., Diep, D., Nissenson, R.A. and Strewler, G.J. High levels of a parathyroid hormone-like protein in milk. Proc. Natl. Acad. Sci. USA, 86, 7183, 1989.

28. Thiede, M.A. The mRNA encoding a parathyroid hormone-like protein is produced in mammary tissue in response to elevations in serum prolactin. Mol. Endocrinol. 3, 1442, 1989.

29. McGrath, C.M., Soule, H.D. Calcium regulation in normal human mammary epithelial cell growth in culture. In Vitro, 20, 652, 1984.

30. Thiede, M.A., Daifotis, A.G., Weir, E.C., Brines, M.L., Burtis, W.J., Ikeda, K., Dreyer, B.E., Garfield, R.E. and Broadus, A.E. Intrauterine occupancy controls expression of parathyroid hormone-related peptide gene in preterm rat myometrium. Proc. Natl. Acad. Sci. USA, 87, 6969, 1990.

PHYSIOLOGICAL AND BIOCHEMICAL ROLES OF THE CHROMOGRANIN FAMILY OF PROTEINS

David V. Cohn and Sven–Ulrik Gorr

Departments of Oral Health and Biochemistry
University of Louisville Health Sciences Center
Louisville, Kentucky, USA 40292

There is increasing evidence that members of the chromogranin family of proteins play important structural and/or modulatory roles in endocrine and neuroendocrine secretory function. The chromogranin family consists of a group of related acidic proteins that are widely distributed in the secretory granules of most endocrine cells studied, including those of the adrenal, pancreas, pituitary, thyroid and parathyroid (1–5). Immunological identification indicates that chromogranin–related proteins are present in secretory granules of fish and unicellular organisms (6–9).

The chromogranin family includes the major proteins chromogranin A (CgA)[1], chromogranin B (CgB), secretogranin II (SgII) and at least three other proteins 7B2, HISL–19 antigen and the 1B1075 gene product. The most studied, and from the physiological standpoint, best understood of these proteins is chromogranin A. For brevity, this review will focus on this protein. Readers may consult a recent review by Huttner et al. (10) for details on the other proteins of the chromogranin family.

Chemistry of CgA -- CgA, depending on mammalian species, contains about 450 amino acids (Figure 1), one–fourth of which are acidic (Glu/Asp). The molecule is O–glycosylated (11,12), phosphorylated (13) and sulfated (3,14). There is substantial conservation of primary structure in the five mammalian species studied to date: human (15,16), rat (17–19), mouse (20), cow (21–23) and pig (24). Particularly noteworthy across these species are 7 conserved regions containing pairs of basic amino acids (Lys–Arg, Arg–Arg,and Lys–Lys) and 20 regions containing either Lys or Arg.

CgA, CgB, and SgII are calcium–binding proteins (25–29), in part the result of the large number of negatively charged amino acids. Of possible importance, a portion of the CgA molecule contains similarity to the calcium–binding region of the E–F hand (16) found in other calcium–binding proteins (30). The binding of calcium by CgA, CgB and SgII leads to aggregation and precipitation from solution (28, 31), see below.

Possible Physiological Roles for the Chromogranins -- The widespread and abundant distribution of CgA and the other chromogranins led investigators early on to consider possible physiological roles for these proteins. Two speculations (see 10,26,32,33) have recently been supported by experimental evidence: First, that they serve as peptide precursors or prohormones. Second, that they play a role in intracellular sorting and packaging of exportable proteins.

---

[1] CgA is also referred to as Secretory Protein I; Chromogranin B was previously termed Secretogranin I, while Secretogranin II was named Chromogranin C.

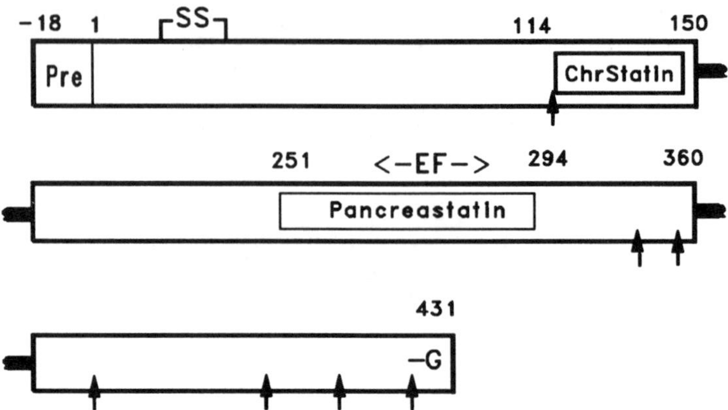

Figure 1. Structural Features of Chromogranin A. The schematic is a composite of the human, cow, pig, rat and mouse proteins. Arrows represent pairs of conserved dibasic cleavage sites. Numbers representing the location of the chromostatin (ChrStatin) and pancreastatin sequences are specifically for bovine CgA and are approximate for the other species. E–F refers to the E–F hand, (see text).

A Prohormone Role for the Chromogranins -- Basic amino acid residues, particularly when in pairs, often serve as cleavage sites in peptides for generation of biologically active moieties (34). Examples of such peptides include proparathormone that upon cleavage yields parathormone and a hexapeptide; and proopiomelanocortin that yields several active peptides including ACTH, ßMSH, and endorphins. The conservation of several potential cleavage sites in CgA formed the basis of the speculation that CgA was a proprotein or prohormone.

Pancreastatin is a 49 amino acid, carboxyl–terminally amidated peptide that was originally isolated from porcine pancreas (35). A similar peptide has also been isolated from bovine pancreas (36). The sequence of pancreastatin is fully replicated in a central region of CgA from the appropriate species. Furthermore, the pancreastatin region is flanked by a Gly residue that would be required for carboxyl–terminal amidation as well as putative basic cleavage sites (though not necessarily as pairs) at the N– and C–termini.

Pancreastatin is a potent inhibitor of glucose–stimulated insulin release by the pancreas (35–37), hypocalcemia–induced parathormone secretion by the parathyroid (38), cholecystokinin–stimulated secretion by the exocrine pancreas (39) and stimulated acid secretion by parietal cells (40). In some cases, pancreastatin stimulates rather than inhibits secretion. For example, it has been reported to enhance pancreatic glucagon secretion (41) and glucagon–stimulated insulin secretion (42).

CgA contains other biologically active sequences. Region 124 to 143 of the bovine molecule corresponds to a 20 amino acid peptide termed chromostatin that has been shown to inhibit secretion of epinephrine by bovine adrenal chromaffin cells in culture (43). In contrast, pancreastatin is ineffective in this system. We may infer from these data that peptides derived from different regions of the parent molecule exhibit unique target cell specificities. CgA, itself, when added to in vitro test systems effectively inhibits stimulated epinephrine (43) and parathormone (44) release. However, higher doses of CgA than of pancreastatin are required for this action or the added CgA must incubate for a time with the target cells. While it is possible that CgA has inherent biological activity, these findings imply that it is processed to an active peptide during incubation with the tissue. Although CgB and SgII contain an even greater number of dibasic sites than CgA, no biologically active peptides based on the former proteins have been identified as yet.

Is the Action of CgA on Secretion of Physiological Importance? Two possibilities suggest themselves: The first is that CgA–derived peptides modulate secretion by the cells from which they are secreted or by the cells in the immediate vicinity. We have speculated that an autocrine/paracrine inhibition of secretion by pancreastatin would sharpen individual hormone pulses, thus providing more subtle control over the amount of hormone released to the circulation (33,38). Conversely, when a CgA–derived peptide stimulates secretion, amplification of the action of the secretory agonist would result. Both conditions could contribute to endocrine homeostasis.

A second possibility is that CgA peptides are endocrine factors that act on distant target cells. The likelihood of this type of control is supported by the report that CgA at a concentration approximating that in the circulation affected insulin release when tested in the perfused rat pancreas (37).

If the latter situation is found to hold, CgA might be playing hitherto unrecognized roles in gland to gland communication. There are some intriguing hints in the literature that seem to us suggestive of such a relationship. For example, in hyperparathyroidism, in which circulating CgA would likely be elevated, a mild hypoinsulin diabetes is occasionally noted. One might postulate that excess CgA released from the parathyroid is repressing insulin release from the islets.

A second example relates to a role for CgA in the regulation of blood pressure. Hypertension is often a manifestation of pheochromocytoma, a syndrome in which circulating CgA is extremely high (45). It has been reported that in human hypertension, circulating CgA is elevated (46). In the stroke–prone hypertensive rat (47), the levels of adrenal and circulating CgA are markedly elevated. Moreover, CgA is costored in the secretory granules of the heart atrium and the adrenal medulla with atrial natriuretic factor and catecholamines, respectively (48,49). In this light, it is possible to see how CgA could play a role in the regulation of blood pressure: normally, atrial CgA could inhibit catecholamine secretion from the adrenal gland. In pheochromocytoma, the large amounts of CgA release to the circulation could inhibit the secretion of atrial natriuretic factor from the myocardial atrium, thus potentiating hypertension.

Does CgA Play a Role in Directing Intracellular Traffic of Exportable Proteins? -- Sorting of regulated secretory proteins, including peptide hormones and chromogranins, involves their specific aggregation in the trans–Golgi compartment. The aggregated proteins are packaged into nascent secretory granules while non–aggregating constitutive secretory proteins are rapidly secreted by small transport vesicles (50). The widespread occurrence of chromogranins in endocrine secretory granules raises the possibility that these proteins play a role in the sorting and packaging of secretory proteins in endocrine cells.

Recent evidence supports such a role for the chromogranins. We, and others, have reported that CgA (26–28), CgB (28) and SgII (31) aggregate at low pH in the presence of millimolar concentrations of calcium. These conditions correspond to those in the trans–Golgi compartments and secretory granules of endocrine cells (51,52). The chromogranin aggregates can include parathormone (28), but exclude the constitutive secretory proteins ovalbumin and immunoglobulin (28,31).

These data have led to a model for sorting and packaging of chromogranins and possibly other regulated secretory proteins (10, 26): According to this model (Figure 2), a high calcium concentration and a low pH, conditions that are found in the sorting organelles of endocrine and exocrine cells, are required for correct sorting and packaging into secretory granules.

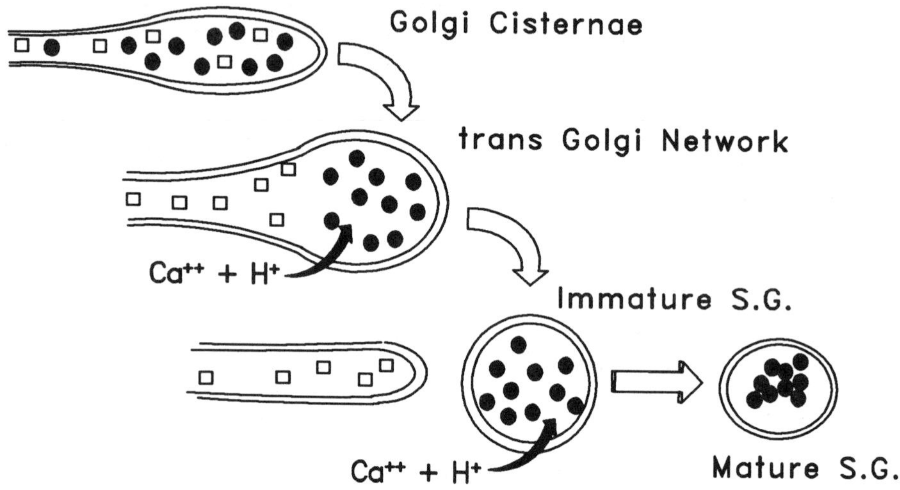

Figure 2. Putative roles of CgA in directing traffic in the secretory pathway of an endocrine cell and in the maturation of secretory granules. CgA (filled circles) and extraneous protein (open squares) move together through the cisternae of the Golgi. In the trans–Golgi zone, Ca^{++} and H^{+} are accumulated. CgA, but not non–regulated proteins, aggregates under these conditions. Other regulated endocrine peptides (e.g., parathormone) may be included in the CgA aggregates. The aggregates are packaged into immature secretory granules (S.G.). Additional concentration of Ca^{++} and H^{+} causes further condensation of CgA during maturation of secretory granules. Adapted from References 10 and 26.

Since a common ionic environment, rather than specific sorting receptors, would be required for this proceess to proceed, this model can readily explain the many examples of successful sorting of foreign proteins in transfected cells (53–55). As the secretory proteins are transported along the secretory route they encounter a progressively more acidic environment and an increased calcium concentration. This would lead to the aggregation of chromogranins and other regulated secretory proteins while constitutive secretory proteins remain soluble.

Individual proteins may co–aggregate or form distinct aggregates according to their aggregation potential along the secretory route. Thus, this model can explain the presence of different secretory granules, each carrying a subset of regulated secretory proteins, in individual cells (56,57). The aggregated proteins would be packaged into secretory granules, possibly aided by calcium–induced membrane interaction as was shown for CgA (27).

An attractive feature of this model is that the protein aggregates would be maintained during storage in secretory granules as endocrine secretory granules contain a high calcium concentration (10–40 mM) and an acidic pH (5.5–6.5). Such aggregation is believed necessary to forestall osmolysis of the granules that might occur due to their high internal protein concentrations.

REFERENCES

1. O'Connor DT. Chromogranin: widespread immunoreactivity in polypeptide hormone producing tissues and in serum. Regul. Pept., 6, 263, 1983.

2. Cohn DV, Elting JJ, Frick M, Elde R. Selective localization of the parathyroid secretory protein–I/adrenal medulla chromogranin A protein family in a wide variety of endocrine cells of the rat. Endocrinology, 114, 1963, 1984.

3. Rosa P, Hille A, Lee RWH, Zanini A, De Camilli P, Huttner WB. Secretogranins I and II: two tyrosine–sulfated secretory proteins common to a variety of cells secreting peptides by the regulated pathway. J. Cell Biol., 101, 1999, 1985.

4. Fischer–Colbrie R, Hagn C, Schober M. Chromogranins A, B, and C: widespread constituents of secretory vesicles. Ann. NY Acad. Sci., 493, 120, 1987.

5. Simon J–P, Aunis D. Biochemistry of the chromogranin A protein family. Biochem. J., 262, 1, 1989.

6. Tisserand–Jochem E–M, Lopez E, Milet C, Vidal B, Magnac C, Eyquem A, Cohn DV. Co–localization and secretion of parathyrin of Stannius corpuscles (immunoreactive parathyroid hormone) and secretory glycoproteins including secretory protein–I in the European eel (Anguilla anguilla L.). Bone Mineral, 2, 163, 1987.

7. Deftos LJ, Bjornsson B Th, Burton DW, O'Connor DT, Copp DH. Chromogranin A is present in and released by fish endocrine tissue. Life Sci 40, 2133, 1987.

8. Rieker S, Fischer–Colbrie R, Eiden L, Winkler H. Phylogenetic distribution of peptides related to chromogranins A and B. J. Neurochem., 50, 1066, 1988.

9. Peterson JB, Nelson DL, Ling E, Angeletti RH. Chromogranin A–like proteins in the secretory granules of a protozoan, Paramecium tetraurelia. J. Biol. Chem., 262, 17264, 1987.

10. Huttner, Gerdes H–H, Rosa P. The granin (chromogranin/ secretogranin) family. Trends Biol. Sci., 16, 27, 1991.

11. Morrissey JJ, Hamilton JW, Cohn DV. The secretion of parathormone and glycosylated proteins by parathyroid cells in culture. Biochem Biophy Res Comm 82, 1279, 1978.

12. Kiang W–L, Krusius T, Finne J, Margolis RU, Margolis RK. Glycoproteins and proteoglycans of the chromaffin granule matrix. J. Biol. Chem., 257, 1651, 1982.

13. Bhargava G, Russell J, Sherwood LM. Phosphorylation of parathyroid secretory protein. Proc. Natl. Acad. Sci. U.S.A., 80, 878, 1983.

14. Gorr S–U, Hamilton JW, Cohn DV. Sulfated secreted forms of bovine and porcine parathyroid chromogranin A (secretory protein I). J. Biol. Chem., 266, 5780, 1991.

15. Konecki DS, Benedum UM, Gerdes H–H, Huttner WB, The primary structure of human chromogranin A and pancreastatin. J. Biol. Chem., 262, 17026, 1987.

16. Helman LJ, Ahn TG, Levine MA, Allison A, Cohen PS, Cooper MJ, Cohn DV, Israel MA. Molecular cloning and primary structure of human chromogranin A (secretory protein I) cDNA. J. Biol. Chem., 263, 11559, 1988.

17. Iacangelo I, Okayama H, Eiden LE. Primary structure of rat chromogranin A and distribution of its mRNA. FEBS Lett., 227, 115, 1988.

18. Hutton JC, Nielsen E, Kastern W. The molecular cloning of the chromogranin A–like precursor of ß–granin and pancreastatin from the endocrine pancreas. FEBS Lett., 236, 269, 1988.

19. Parmer RJ, Koop AH, Handa MT, O'Connor DT. Molecular cloning of chromogranin A from rat pheochromocytoma cells. Hypertension, 14, 435, 1989.

20. O'Connor DT et al. J. Biol.Chem., 1991, in press

21. Benedum UM, Baeuerle PA, Konecki DS, Frank R, Powell J, Mallet J, Huttner WB. The primary structure of bovine chromogranin A: a representative of a class of acidic secretory proteins common to a variety of peptidergic cells. EMBO J., 5, 1495, 1986.

22. Iacangelo I, Affolter H–U, Eiden LE, Herbert E, Grimes M. Bovine chromogranin A sequence and distribution of its messenger RNA in endocrine tissues. Nature., 323, 82, 1986.

23. Ahn TG, Cohn DV, Gorr SU, Ornstein DL, Kashdan MA, Levine MA. Primary structure of bovine pituitary secretory proteins I (chromogranin A) deduced from the cDNA sequence. Proc. Natl. Acad. Sci. U.S.A., 84, 5043, 1987.

24. Iacangelo AL, Fischer–Colbrie R, Koller K, Brownstein MJ, Eiden LE. The sequence of porcine chromogranin A messenger RNA demonstrates chromogranin A can serve as the precusor for the biologically active hormone, pancreastatin. Endocrinology, 122, 2339, 1988b.

25. Reiffen FU, Gratzl M. Chromogranins, widespread in endocrine and nervous tissue, bind Ca^{2+}. FEBS Lett., 195, 327, 1986.

26. Gorr S–U, Dean WL, Kumarasamy R, Cohn DV. Calcium binding properties of parathyroid secretory protein–I, in Calcium regulation and bone metabolism: basic and clinical aspects, 9, Cohn DV, Martin TJ, Meunier PJ, Eds. Elsevier, Amsterdam, 1987, 49.

27. Gorr S–U, Dean WL, Radley TL, Cohn DV. Calcium–binding and aggregation properties of parathyroid secretory protein–I (chromogranin A). Bone Mineral, 4, 17, 1988.

28. Gorr S–U, Shioi J, Cohn DV. Interaction of calcium with porcine adrenal chromogranin A (secretory protein–I) and chromogranin B (secretogranin B). Am. J. Physiol., 257, E247, 1989.

29. Cozzi MG, Zanini A. Secretogranin II is a Ca^{2+}–binding protein. Cell Biol. Int. Rep., 12, 493, 1988.

30. Kretsinger RH. Structure and evolution of calcium–modulated proteins. Critical Reviews in Biochemistry, July, 119, 1980.

31. Gerdes H–H, Rosa P, Phillips E, Baeuerle P, Frank R, Argos P, Huttner WB. The primary structure of secretogranin II, a widespread tyrosine–sulfated secretory granule protein that exhibits low pH– and calcium–induced aggregation. J. Biol. Chem., 264, 12009, 1989.

32. Gorr S–U, Kumarasamy R, Dean WL, Cohn DV. New suggestions for the physiological role of secretory protein–I. Bone Mineral, 2, 251, 1987.

33. Cohn DV, Fasciotto BH, Gorr S–U, Parkins F, Shioi J, Levine MA, Greeley GH Jr. The putative role of secretory protein–I / chromogranin A as a precursor for regulatory hormones. Mol. Cell. Regul. Calcium Phosphate Metabol., Peterlik M, Bronner F, Eds. Alan R. Liss, Inc, New York, 1990, 51.

34. Docherty K, Steiner DF. Post–translational proteolysis in polypeptide hormone biosynthesis. Ann. Rev. Physiol., 44, 625, 1982.

35. Tatemoto K, Efendic S, Mutt V, Makk G, Feistner GJ, Barchas JD. Pancreastatin, a novel pancreatic peptide that inhibits insulin secretion. Nature, 324, 476, 1986.

36. Nakano I, Funakoshi A, Miyasaka K, Ishida K, Makk G, Angwin P, Chang D, Tatemoto K. Isolation and characterization of bovine pancreastatin. Regul. Pept., 25, 207, 1989.

37. Greeley GH Jr., Thompson JC, Ishizuka J, Cooper CW, Levine MA, Gorr S–U, Cohn DV. Inhibition of glucose–stimulated insulin relaese in the perfused rat pancreas by parathyroid secretory protein–I (chromogranin–A). Endocrinology, 124, 1235, 1989.

38. Fasciotto BH, Gorr S–U, DeFranco DJ, Levine MA, Cohn DV. Pancreastatin, a presumed product of chromogranin–A (secretory protein–I) processing, inhibits secretion from porcine parathyroid cells in culture. Endocrinology, 125, 1617, 1989.

39. Funakoshi A, Miyasaka K, Nakamura R, Kitani K, Funakoshi S, Tamamura H, Fujii N, Yajima H. Bioactivity of synthetic human pancreastatin on exocrine pancreas. Biochem. Biophys. Res. Commun., 156, 1237, 1988.

40. Lewis JJ, Zdon MJ, Adrian TE, Modlin IM. Pancreastatin: a novel peptide inhibitor of parietal cell secretion. Surgery, 104, 1031, 1988.

41. Ahrén B, Lindskog S, Tatemoto K, Efendic S. Pancreastatin inhibits insulin secretion and stimulates glucagon secretion in mice. Diabetes, 37, 281, 1988.

42. Ishizuka J, Tatemoto K, Cohn DV, Thompson JC, Greeley GH. Manuscript in preparation.

43. Galindo E, Rill A, Bader M–F, Aunis D. Chromostatin, a 20–amino acid peptide derived from chromogranin A, inhibits chromaffin cell secretion. Proc. Natl. Acad. Sci. U.S.A., 88, 1426, 1991.

44. Fasciotto BH, Gorr S–U, Bourdeau AM, Cohn DV. Autocrine regulation of parathyroid secretion: inhibition of secretiuon by chromoganin–A (secretory protein–I) and potentiation of secretion by chromogranin–A and pancreastatin antibodies. Endocrinology, 127, 1329, 1990.

45. O'Connor DT, Deftos LJ. Secretion of chromogranin A by peptide–producing endocrine neoplasms. N. Engl. J. Med., 314, 1145, 1986.

46. Takiyyuddin MA, Cervenka JH, Hsiao RJ, Barbosa JA, Parmer RJ, O'Connor DT. Chromogranin A. Storage and release in hypertension. Hypertension, 15, 237, 1990.

47. Schober M, Howe PRC, Sperk G, Fischer–Colbrie R, Winkler H. An increased pool of secretory hormones and peptides in adrenal medulla of stroke–prone spontaneously hypertensive rats. Hypertension, 13, 469, 1989.

48. Steiner H–J, Weiler R, Ludescher C, Schmid KW, Winkler H. Chromogranins A and B are co–localized with atrial natriuretic peptides in secretory granules of rat heart. J. Histochem. Cytochem., 38, 845, 1990.

49. Winkler H, Sietzen M, Schober M. The life cycle of catecholamine–storing vesicles. Ann. NY Acad. Sci., 493, 3, 1987.

50. Kelly RB. Pathways of protein secretion in eukaryotes. Science, 230, 25, 1985.

51. Bulenda D, Gratzl M. Matrix free Ca^{2+} in isolated chromaffin vesicles. Biochemistry, 24, 7760, 1985.

52. Orci L, Ravazzola M, Amherdt M, Madsen O, Perrelet A, Vassalli J–D, Anderson RGW. Conversion of proinsulin to insulin occurs coordinately with acidification of maturing secretory vesicles. J. Cell Biol., 103, 2273, 1986.

53. Burgess TL, Craik CS, Kelly RB. The exocrine protein trypsinogen is targeted into the secretory granules of an endocrine cell line: studies by gene transfer. J. Cell Biol., 101, 639, 1985.

54. Moore H–pH, Kelly RB. Re–routing of a secretory protein by fusion with human growth hormone sequences. Nature, 321, 443, 1986.

55. Powell SK, Orci L, Craik CS, Moore H–PH. Efficient targeting to storage granules of human proinsulins with altered propeptide domain. J. Cell Biol., 106, 1843, 1988.

56. Hashimoto S, Fumagalli G, Zanini A, Meldolesi J. Sorting of three secretory proteins to distinct secretory granules in acidophilic cells of cow anterior pituitary. J. Cell Biol., 105, 1579, 1987.

57. Wagner DD, Saffaripour S, Bonfanti R, Sadler JE, Cramer EM, Chapman B, Mayadas TN. Induction of specific storage organelles by von Willebrand factor propolypeptide. Cell, 64, 403, 1991.

TRIIODOTHYRONINE AND 1,25-DIHYDROXYVITAMIN D_3: ACTIONS AND INTERACTIONS AT THE GENE, CELL AND ORGAN LEVEL

Heide S. Cross, Klaus Klaushofer and Meinrad Peterlik
Department of General and Experimental Pathology, University of Vienna Medical School, and Ludwig Boltzmann Research Unit for Experimental and Clinical Osteology, Hanusch Hospital, Vienna, Austria

The main reason for considering thyroxine (T_4) and triiodothyronine (T_3) as calcium and phosphate regulating hormones stems from the fact that diseases of the thyroid gland are inevitably associated with disturbances of calcium and inorganic phosphate (Pi) homeostasis (for review, see ref. 1). Thus, thyroid patients present with alterations in plasma calcium and Pi levels, changes in the rate of intestinal absorption and renal excretion of calcium and Pi, which, accordingly, affect total body mineral ion balance and eventually become manifest as metabolic bone disease. For example, untreated hyperthyroidism is accompanied by accelerated bone loss leading to metabolic osteoporosis, while, conversely, bone density in patients with hypothyroidism in general is higher than normal.[2-5]

This leads to the question how thyroid hormones are implicated in the regulation of mineral ion metabolism. In this paper we wish to summarize the available evidence to indicate that thyroid hormones, apart from modulating the biosynthesis of 1,25-dihydroxyvitamin D_3 (1,25$(OH)_2D_3$), have direct vitamin D-like effects on intestinal, renal and bone cells, and, in addition, are able to potentiate the cellular effects of the steroid hormone on calcium and Pi movements in its classical target organs.

I. 1,25$(OH)_2D_3$ AND THYROID HORMONES: MUTUAL INFLUENCE ON BIOSYNTHESIS

Low levels of circulating 1,25$(OH)_2D_3$ are frequently observed in hyperthyroidism and have been suggested to result from an indirect action of thyroid hormones on renal metabolism of 25-hydroxyvitamin D: Through its calcium-mobilizing action on bone, T_3 would be able to raise serum calcium and thereby to suppress secretion of PTH from the parathyroid gland. This in turn would lower the activity of the 1α-25-hydroxyvitamin D-hydroxylase in the kidney.[6] Recently, Bouillon et al.[7] put forward an additional explanation for reduced 1,25$(OH)_2D_3$ synthesis in the hyperthyroid state by showing that thyroid hormones can directly inhibit the 1α-hydroxylase activity.

While this constitutes rather clear evidence for a regulatory role of thyroid hormones in the biosynthesis of the active vitamin D sterol, little is known whether and how $1,25(OH)_2D_3$ *vice versa* is engaged in mechanisms controlling synthesis and secretion of thyroid hormones. Stumpf[8] has provided distinct evidence for the presence of high affinity vitamin D binding molecules, apart from brain TRH neurones, in pituitary thyreotroph as well as in thyroidal follicular cells. Therefore, both cell types must be considered targets for $1,25(OH)_2D_3$ and it is thus entirely possible that $1,25(OH)_2D_3$ is involved in the regulation of thyroid hormone production along the hypothalamic/pituitary/thyroid gland axis.[9] This assumption is supported by the observation that the sterol, although it does not change basal TSH production in primary cultured rat pituitary cells, it augments TSH release from these cells in response to TRH.[10] Since this effect of $1,25(OH)_2D_3$ was observed only at a relatively high concentration in the culture medium of 10^{-8} M, its relevance for a stimulatory action of $1,25(OH)_2D_3$ on thyroid hormone production is not clear as yet. In contrast, at the thyroid gland level, $1,25(OH)_2D_3$ rather seems to attenuate basal and TSH-mediated cell functions. In the rat thyroidal cell line FRTL-5, $1,25(OH)_2D_3$ dose-dependently inhibited iodide uptake and cAMP accumulation, although an actual effect on thyroid hormone production has not been reported).[11] Therefore, it remains to be seen whether 1,25(OH)2D3 is able to attenuate thyroidal functions and, consequently, hormone production also *in vivo*. If so, this effect could modulate the efficiency of the steroid hormone´s action on TSH production by pituitary cells.

II. $1,25(OH)_2D_3$ AND THYROID HORMONE ACTIONS ON THE SMALL INTESTINE

Underlying the regulatory role of vitamin D in mineral ion and skeletal homeostasis are the well known actions of its hormonal form, $1,25(OH)_2D_3$, on intestinal absorptive processes of calcium and Pi. For an up-date on this subject see ref. 12.

The exact mechanism by which the sterol exerts control over intestinal absorption is not yet fully understood, although numerous studies have been carried out to show that each single step constituting the transcellular pathway of calcium transfer, i.e., entry across the brush-border membrane, transcellular migration and basolateral extrusion, is sensitive to vitamin D. The most likely model to explain vitamin D actions was derived from experimental evidence as well as from theoretical considerations by Bronner.[13] It implies the vitamin D-induced calcium-binding protein, calbindin, as rate-limiting factor of $1,25(OH)_2D_3$-dependent transcellular calcium transport. It is now well established that vitamin D bound to its receptor transactivates the calbindin gene, and, in parallel with the extent of its expression, induces transcellular calcium transfer.

The role of thyroid hormones on intestinal calcium absorption has been investigated utilizing organ-cultured small intestinal segments from 20 day old embryonic chicks. Neither T_4 nor T_3 had any significant effect on basal calcium transport as measured after 48 h culture in a vitamin D-free medium. In the presence of vitamin D, both thyroid hormones augmented the effect of the sterol on calbindin synthesis[14] as well as on calcium uptake.[15-17] The potentiating effect of T_3 on $1,25(OH)_2D_3$ action can be explained by the ability of T_3 to act as an auxiliary transcriptional factor for the calbindin gene.

Since the effect of $1,25(OH)_2D_3$ on intestinal calcium transport has been linked to receptor-mediated changes of calbindin gene expression,we conclude that the permissive effect of T_3 on $1,25(OH)_2D_3$-induced calcium transport probably occurs during the multi-step process of activation and control of gene expression. Although it cannot be excluded that T_3 increases the number of vitamin D receptors (VDR), an interaction of the two hormones at the DNA level has to be considered a likely possibility. A logical basis for this assumption is provided by the high degree of sequence homology between the nuclear VDR and the T_3 receptor.[18] Steroid hormone-sensitive genes apparently contain clusters of overlapping "hormone response elements" in their regulatory region, and can therefore be activated in a cooperative fashion by receptor mediated binding of one hormone in combination with another member of the steroid/thyroid hormone family.[19] In this respect it must be noted that the rat intestinal calbindin gene promotor contains multiple hormone response elements which could convey sensitivity to the gene towards, apart from $1,25(OH)_2D_3$, also other members of the hormone family like thyroid hormone.[20]

Absorption of Pi from the lumen of the small intestine is subject to regulation by vitamin D through its stimulatory effect on Na^+ gradient-driven Pi transfer across the brush border membrane of enterocytes (for review, see among others, refs. 21-23). The effectiveness of vitamin D has been shown *in vivo* by administration of $1,25(OH)_2D_3$ to vitamin D-deficient chicks as well as *in vitro*, e.g., by addition of the steroid hormone to organ-cultured explants of embryonic chick small intestine. In both instances, $1,25(OH)_2D_3$ brought about a two to threefold increase in the intrinsic activity of the Na^+/Pi cotransport system at the brush border membrane which could be attributed mainly to a corresponding increase in the number of carrier sites. In addition, through an appropriate effect on transmembrane Na^+ fluxes (through potential-sensitive channels and via molecularly coupled Na^+/H^+ exchange)[24], $1,25(OH)_2D_3$ makes more energy available from the transmembrane Na^+ gradient for Pi transfer against an electrochemical gradient.

A vitamin D-like effect of T_3 on Na^+/Pi transport across the brush-border membrane of electrocytes could be demonstrated in organ-cultured embryonic chick jejunum. The thyroid hormone enhances Pi uptake by explants from 20 day old embryos even more efficiently than $1,25(OH)_2D_3$. This effect of T_3 showed a linear concentration dependence between 10^{-10} - 10^{-7} M and was sensitive to inhibition of protein synthesis, e.g. by cycloheximide. This strongly suggests a genomic effect of T_3 underlying the observed increase in Na^+-dependent Pi transfer.[15] In brush-border membrane vesicles isolated from organ-cultured embryonic chick jejunal explants, T_3, like $1,25(OH)_2D_3$, enhanced intravesicular Pi accumulation driven by an outside/inside Na+ gradient (100/0 mM). Since T_3 had no concomitant effect on Na^+ fluxes and thus on energy expenditure from the Na^+ gradient, this implies that the thyroid hormone alters the intrinsic activity of the Na^+/Pi carrier system.[25] It is conceivable that this reflects enhanced synthesis of Na^+/Pi carrier molecules, since expression of this T_3 effect, as mentioned before, requires intact protein synthesis.

This raises the question whether a single gene codes for Na^+/Pi transport which can be activated by $1,25(OH)_2D_3$ and T_3 alike, or whether enterocytes express two carrier systems coded for by genes with differential sensitivity to the steroid and the thyroid hormone. A partial answer to this question can be derived from the observation that, apart from its direct action on the Na^+ gradient-driven Pi transport system of the small intestinal brush-border membrane, T_3 apparently is able to additionally enhance intestinal Pi absorption through interaction with $1,25(OH)_2D_3$. When both hormones were added together to cultures of embryonic chick jejunal segments a more than additive effect on cellular Na^+-dependent Pi uptake was observed.[16, 17] Utilizinq brush-border membrane vesicles from cultured jejuna, it became apparent, that the synergistic action of the two hormones, which is likely to occur at the gene level, is expressed as an increment in Na^+ gradient driven Pi membrane transport.[25] Analysis of the dose-response relationships of the thyroid/steroid hormone interaction suggests that $1,25(OH)_2D_3$ has no influence on gene activation by T_3, while the thyroid hormone permissively enhances the genomic effect of $1,25(OH)_2D_3$.[16]

An interaction of T_3 with $1,25(OH)_2D_3$ became also apparent, when regulation of Na^+/D-glucose transport by the two hormones was studied in isolated brush-border membrane vesicles (BBMV) from embryonic chick jejunum.[26] When gut explants were cultured in the presence of T_3 (10^{-8} M) alone, no change in D-glucose uptake by BBMV could be demonstrated. In contrast, $1,25(OH)_2D_3$ (at 10^{-8} M) had a pronounced effect on the extent of Na^+-gradient driven intravesicular accumulation of D-glucose ("overshoot" phenomenon). When D-glucose transfer was also determined under Na^+ equilibrium as well as voltage-clamp conditions to exclude any simultaneous effect of the steroid hormone on the electrogenic transport system through changes in ionic permeabilities of the brush-border membrane, it became clear that also in the case of Na^+/Pi-glucose transport, $1,25(OH)_2D_3$ increases the intrinsic activity of the carrier system. The corresponding rise in the number of carrier sites is reflected by the appearance of additional high affinity phlorizine binding sites and also by an approximately twofold rise of the maximal velocity of the carrier system. A further increase in V_{max} of D-glucose uptake by BBMV was observed when T_3 was added simultaneously with $1,25(OH)_2D_3$ to cultures of jejunal explants.

III. $1,25(OH)_2D_3$ AND T_3 ACTIONS ON RENAL EPITHELIAL CELLS

$1,25(OH)_2D_3$ is certainly one of many important factors that regulate the reabsorption of calcium from the tubular fluid. In man like in many other species, the influence of the steroid hormone on calcium reabsorption is most prominent in the distal convoluted tubule. Again, like in the small intestine, the action of $1,25(OH)_2D_3$ is mediated through appropriate changes in the expression of the calbindin gene.[27]

To the best of our knowledge, no studies have been carried out so far on a possible effect of thyroid hormones on calcium reabsorption in the kidney. However, under the assumption that the renal calbindin gene contains the same array of hormone responsive elements in its 5′ flanking region like the corresponding gene in enterocytes, it seems predictable that thyroid hormones could also modulate $1,25(OH)_2D_3$-related calcium reabsorption across the renal tubular epithelium.

It is a well established fact that Na^+ gradient-driven Pi transport across the renal proximal tubular epithelium responds to certain steroid hormones, e.g. $1,25(OH)_2D_3$ and glucocorticoids. While the latter inhibit Pi uptake, e.g., in cultured renal cells,[28] $1,25(OH)_2D_3$ acts as a stimulator of the Na^+/Pi cotransport system at the brush-border membrane of proximal tubular cells (cf. ref. 22).

Only a single study was done on the effect of thyroid hormones on renal phosphate transport. Espinosa et al.[29] showed that administration of T_4 to rats significantly increased the Vmax of Na^+ gradient driven Pi uptake by brush-border membrane vesicles from renal cortex.

Whether thyroid hormones can interact with $1,25(OH)_2D_3$-related renal Pi absorption, as observed in the small intestine, remains to be seen. Although Pi is transferred via a Na^+ cotransport mechanism across the luminal membrane of renal tubular cells, the renal carrier system differs in a number of important aspects from the intestinal carrier. These include, among others, sensitivity to PTH, modulation by cyclic nucleotides, preference for HPO_4^{2-} over $H_2PO_4^-$ etc. (cf. ref. 21). In view of the fact, that Na^+/Pi cotransport in the kidney is sensitive to a number of members of the steroid/thyroid hormone family, including glucocorticoids (see before), interaction at the level of the genome between, e.g., T_3 and 1,25$(OH)_2D_3$ is thus conceivable.

IV. ACTION OF 1,25$(OH)_2D_3$ AND THYROID HORMONES ON BONE

The essentiality of vitamin D as well as of thyroid hormones for normal skeletal growth and mineralization is most apparent through the hallmarks of the respective hormone deficiency syndromes: Congenital dysfunction of the thyroid gland, if remaining undetected, leads to severe retardation of skeletal growth, whereas in rickets or osteomalacia, vitamin D deficiency becomes most prominent by insufficient mineralization of the organic bone matrix. While the significance of either hormone for the process of bone *modeling* has been recognized since decades, a concept involving 1,25$(OH)_2D_3$ and particularly thyroid hormones in bone *remodeling* or *turnover* is only slowly emerging.

Formation of new bone cannot be accomplished merely by proliferation of osteoblasts but requires coordination of cell growth with development of differentiated cell functions.[30] In this respect it must be emphasized that 1,25$(OH)_2D_3$ is certainly one important factor for the temporal regulation of osteoblast differentiation. In this respect, the sterol stimulates in a differential manner the synthesis of type I collagen, the activity of alkaline phosphatase and notably the secretion of osteocalcin. 1,25$(OH)_2D_3$ thus seems to be able to coordinate osteoblast growth and function to achieve synthesis of bone matrix and other proteins in sufficient amounts and quality, which is a precondition for initiation of the mineralization process.

Although the role of the non-collagenous protein osteocalcin in bone formation and mineralization is not clear at all, its secretion by osteoblasts is considered a measure of the rate of bone formation. Analysis of the regulatory 5' flanking region of the osteocalcin gene has revealed the existence of binding sites for the steroid/thyroid hormone receptor family.[31] While there are numerous reports showing that 1,25$(OH)_2D_3$ is a potent stimulator of osteocalcin production, conflicting results were reported on the effect of T_3 thereon. Rizzoli et al.[32] showed enhancement of osteocalcin synthesis by T_3 in human osteosarcoma cells, while according to Mäenpää et al.[33] the osteocalcin gene can be activated by 1,25$(OH)_2D_3$, but a number of other steroid hormones and particularly T_3 were without any effect on gene expression. However, when added together with 1,25$(OH)_2D_3$, the thyroid hormone diminished the stimulatory effect of the sterol on osteocalcin synthesis.

Even if the mode of interaction between 1,25$(OH)_2D_3$ and T_3 in controlling bone formation needs clarification, there is agreement that both hormones are agonists with respect to their action on bone resorption. Direct stimulation of bone resorption by thyroid hormones was observed in cultures of fetal rat limb bones[6, 34] and of neonatal mouse calvaria.[35, 36] In the latter system, T_3 between 10^{-9}–10^{-7} M dose-dependently enhanced osteoclastic resorption of cultured bone. Simultaneously, the thyroid hormone caused the release of bone resorbing prostaglandins, viz. PGE_2 and PGI_2, into the culture medium. This indicates that the bone resorbing activity of T_3 is at least in part due to the endogenous generation of these prostanoids. This assumption is supported by observations that thyroid hormone-induced resorption of cultured parietal bone is largely reduced when generation of endogenous prostaglandins is suppressed by either indomethacin or γ-interferon.[35]

The variability of the prostaglandin-dependent fraction of thyroid hormone-induced bone resorption suggests that T_3 acts indirectly, perhaps through the release of local growth factors or cytokines (see also ref. 37). This is not unlikely since the bone resorbing potency of a number of such factors like interleukin 1, epidermal growth factor, TNF-α or TGF-β, among others, has been linked to their ability to enhance endogenous prostaglandin (PG) production. Positive evidence for this assumption was obtained by showing that the effect of thyroid hormones on prostaglandin-mediated bone resorption in neonatal mouse calvaria can be blocked by a monoclonal anti-TGF-β-antibody.[38] With respect to the involvement of cytokines and prostaglandins, the bone resorbing activity of thyroid hormones closely resembles that of $1,25(OH)_2D_3$. This assumption is supported by an earlier report from our laboratory that $1,25(OH)_2D_3$-stimulated bone resorption in neonatal mouse calvaria is partially sensitive to cyclooxygenase inhibition by indomethacin or γ-interferon.[39]

The sterol has been shown to enhance the synthesis of prostaglandins preferentially of the E_2-type in different types of human cells such as amnion cells,[40] promyelocytic leukemia (HL-60) cells,[41] peripheral blood lymphocytes,[42] and, notably, monocytes.[43] These findings suggest that the steroid hormone could possibly elicit the release of PGs also from other cells of the monocyte type, e.g. in bone tissue, where cells related to the monocyte/macrophage lineage exist as precursors for osteoclasts.[44] If so, such an effect could certainly contribute to the bone resorbing activity of the sterol. However, the possibility has to be taken into consideration that $1,25(OH)_2D_3$ could also inhibit PG formation in bone tissue. Although the sterol had no influence at all on PGE_2 production by cultured osteoblastic osteosarcoma cells,[45] $1,25(OH)_2D_3$ has been shown to suppress the release of PGE_2 from cultured osteoblast-like cells derived from human bone[46] or from embryonic chick calvaria.[47]

We therefore measured the release of certain PGs with known bone resorbing activity, viz. PGE_2 and PGI_2 (determined as 6-keto-$PGF_{1\alpha}$),[48] from neonatal mouse calvaria cultured in the presence of $1,25(OH)_2D_3$, and also assessed the influence of inhibition of PG synthesis by indomethacin on $1,25(OH)_2D_3$-induced calcium release into the culture medium.

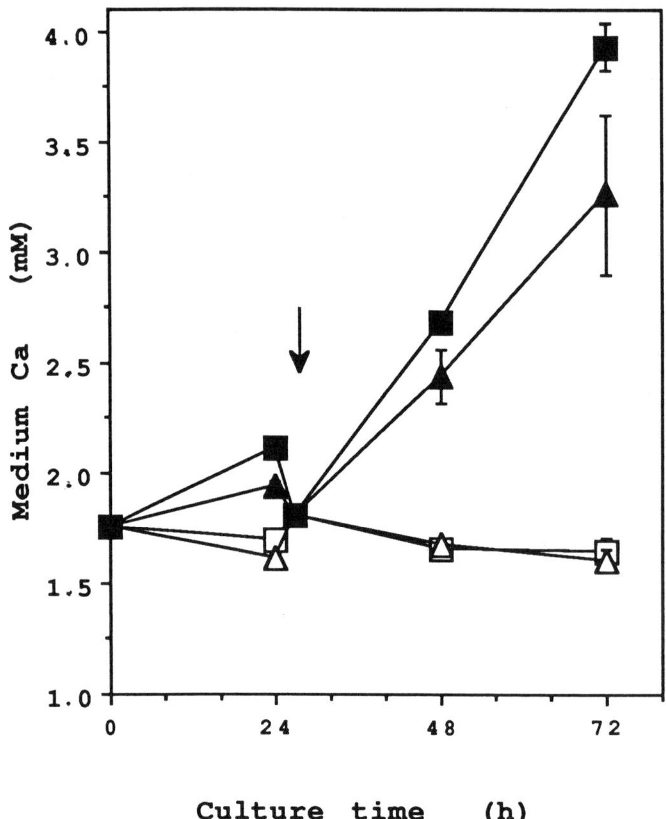

Figure 1: Effect of $1,25(OH)_2D_3$ on calcium release from cultured neonatal mouse calvaria. Bones were cultured as detailed in ref. 49. ❑, no treatment; ■, $1,25(OH)_2D_3$, 1×10^{-9} M; Δ, indomethacin, 5×10^{-7} M; ▲, $1,25(OH)_2D_3$, 1×10^{-9} M, + indomethacin, 5×10^{-7} M. Arrow indicates change of culture medium. Data from a single experiment are expressed as mean ± SEM (from 5 bones per group).

Figure 1 shows the time course of $1,25(OH)_2D_3$-dependent bone resorption. A steady increase in medium calcium under the influence of the hormone indicated ongoing osteoclastic bone resorption within the 72 h time frame. Calcium release from cultured bones could be reduced by indomethacin ($5x10^{-7}$ M) by approximately 30 %.

In a series of experiments the dose-response relation of $1,25(OH)_2D_3$ on prostaglandin-mediated bone resorption was evaluated by comparing the effect of graded hormone concentrations on bones cultured with and without indomethacin. Data collated in Figure 2 show that inhibition of cyclooxygenase activity resulted in significant inhibition of sterol-stimulated calcium release between 10^{-11} - 10^{-9} M $1,25(OH)_2D_3$.

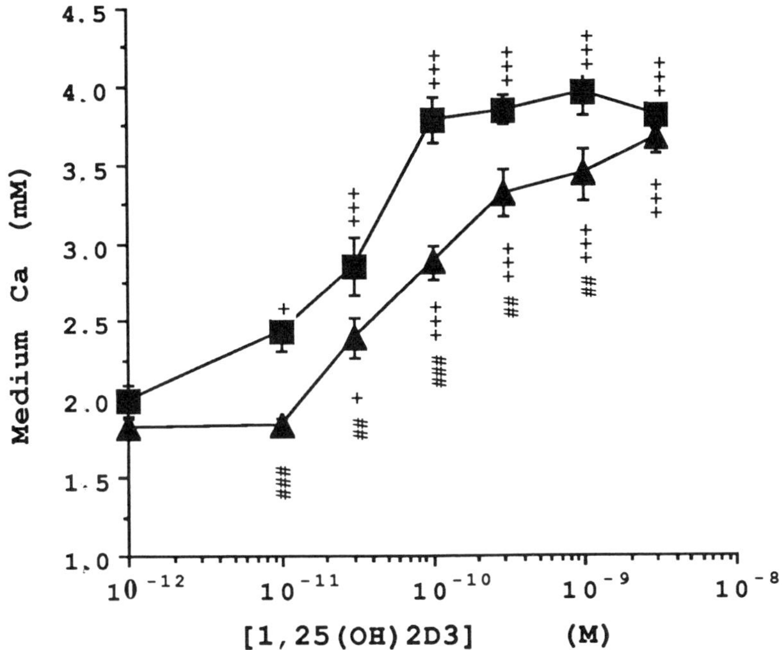

Figure 2: Dose-response relation of $1,25(OH)_2D_3$ on PG mediation of bone resorption. ■, $1,25(OH)_2D_3$; ▲, $1,25(OH)_2D_3$ + indomethacin, $5x10^{-7}$ M. Bones were cultured as detailed in ref. 50. Culture time was 72 h. Data are means ± SEM (n = 15) from three separate experiments. Significance of differences (analysis of variance): + $p < 0.05$, +++ $p < 0.001$ (treated vs. untreated group); ## $p < 0.01$, ### $p < 0.001$ (1,25(OH)2D3 vs. 1,25(OH)2D3 + indomethacin treatment).

The dose-response relation of 1,25(OH)$_2$D$_3$ on PG release (Table 1) showed that the hormone raised PGE$_2$ levels between 10^{-10} - 10^{-9} M, with a maximal effect at 3×10^{-10} M. At the same concentration, a highly significant increase of 6-keto-PGF$_{1\alpha}$ in the culture medium was also observed.

TABLE 1

Effect of 1,25(OH)$_2$D$_3$ on Bone Resorption and Prostaglandin Production in Cultured Calvarial Bones[a]

1,25(OH)$_2$D$_3$ (M)	**Medium Ca^{++}** (mM)		**PGE$_2$** (ng/ml)		**6-keto-PGF$_{1\alpha}$** (ng/ml)	
	-indo	+indo	-indo	+indo	-indo	+indo
none	1.77±0.03	1.67±0.03	0.5±0.11	0.29±0.01	<0.01	<0.01
1x10-11	2.24±0.21	1.66±0.03	0.39±0.08	0.28±0.02	0.16±0.09	<0.01
3x10-11	2.59±0.22***	1.96±0.07	0.55±0.14	0.34±0.04	0.35±0.21	<0.01
1x10-10	3.35±0.22***	2.45±0.14	0.81±0.2*	0.22±0.05	0.53±0.25**	<0.01
3x10-10	3.75±0.14***	2.57±0.14	1.50±0.13***	0.29±0.06	1.21±0.2***	<0.01
1x10-9	3.25±0.21***	3.00±0.24	0.83±0.16*	0.37±0.02	0.27±0.15	<0.01
3x10-9	3.46±0.13***	3.37±0.23	1.06±0.12***	0.36±0.04	0.20±0.11	<0.01

[a] Mouse calvarial bones were cultured as detailed in ref. 50. Culture time 72 h. Calcium concentration of culture medium was 1.80 mM. Indomethacin concentration was 5×10^{-7} M (+indo). Detection limit for PG analysis was 0.01 ng/ml. Data are means ± SEM (n = 5). Significant differences from control group: *$p < 0.05$, **$p < 0.01$, ***$p < 0.001$.

Although determination of PG medium concentrations revealed considerable variations of basal PG release as discussed previously,[51] it was thus apparent that osteolytic prostaglandins tended to accumulate in cultures of 1,25(OH)$_2$D$_3$-stimulated calvariae (Table 1). Since PGs, after release from cultured bone, could undergo further metabolism to a certain extent, their concentrations measured at the end of the culture period might not entirely reflect their preceding resorptive action. We had therefore tried to assess the significance of PG formation in 1,25(OH)$_2$D$_3$-stimulated bone resorption through inhibition of the cyclooxygenase reaction. Particularly the difference in the dose-response relation, which was observed when the effect of 1,25(OH)$_2$D$_3$ was evaluated in the presence of indomethacin (Figure 2), strongly suggests that a certain fraction of the activity of the steroid hormone in bone is due to increased generation of PGs. From these experiments it is also apparent that 1,25(OH)$_2$D$_3$, particularly at low concentrations ($1x10^{-11}$ - $3x10^{-10}$ M), has a pronounced effect on PG formation, while, with increasing hormone concentrations, the contribution of PGs to the extent of 1,25(OH)$_2$D$_3$-mediated resorption becomes less significant (Figure 2). It should be noted that for example bone resorption induced by 1x 10^{-10} M 1,25(OH)$_2$D$_3$ was inhibitable by indomethacin by more than 40 % (Figure 2).

Modulation of membrane phospholipid metabolism by 1,25(OH)$_2$D$_3$ could be one explanation for the effect of the hormone on endogenous PG formation. 1,25(OH)$_2$D$_3$ treatment was shown to change the phospholipid composition of the plasma membrane of various cell types, among them also bone cells.[52] It is interesting to note, that stimulation of phospholipase A_2 activity by 1,25(OH)$_2$D$_3$ apparently depends on the degree of differentiation of a certain cell type, since e.g. in rat costochondral cartilage, 1,25(OH)$_2$D$_3$ activates phospholipase A_2 exclusively in cells derived from the growth zone but not from the resting zone.[53] It seems thus possible, that 1,25(OH)$_2$D$_3$ acts as a stimulator of PG production not only in monocytes/macrophages or osteoclast precursors as it does in monocytes,[43] but also in actively bone forming osteoblasts comparable to growth zone chondrocytes.

Apart from a direct action on phospholipase A_2 activity and subsequent PG formation, 1,25(OH)$_2$D$_3$ could also enhance PG synthesis by indirect means, e.g. through production of bone resorbing cytokines or growth factors.[54] In particular, interleukin 1 (IL-1), TGF-ß or TNF-α, which are potent stimulators of PG synthesis in cultured neonatal mouse calvaria[55-57] are produced under the influence of 1,25(OH)$_2$D$_3$ by various cell types present in bone tissue.[58, 59]

Considering the possibility that 1,25(OH)$_2$D$_3$ as well as T_3 enhance bone resorption through similar if not identical mechanisms of action, we studied resorption of cultured neonatal mouse parietal bone exposed to a combination of both hormones. Figure 3 illustrates an impressive example of the synergistic effect of 1,25(OH)$_2$D$_3$ and T_3 on osteoclastic bone resorption, which, notably, is most prominent at low concentrations of each hormone in the culture medium.

For reasons that were discussed in details elswhere,[16] we believe that the effects of 1,25(OH)$_2$D$_3$ and T_3 observed in culture resemble actions of the two hormones *in vivo*. This implies that their interaction has a profound influence on the constantly ongoing process of bone remodeling and also emphasize the importance of the vitamin D/thyroid axis in maintaining normal structure and function of the skeletal system.

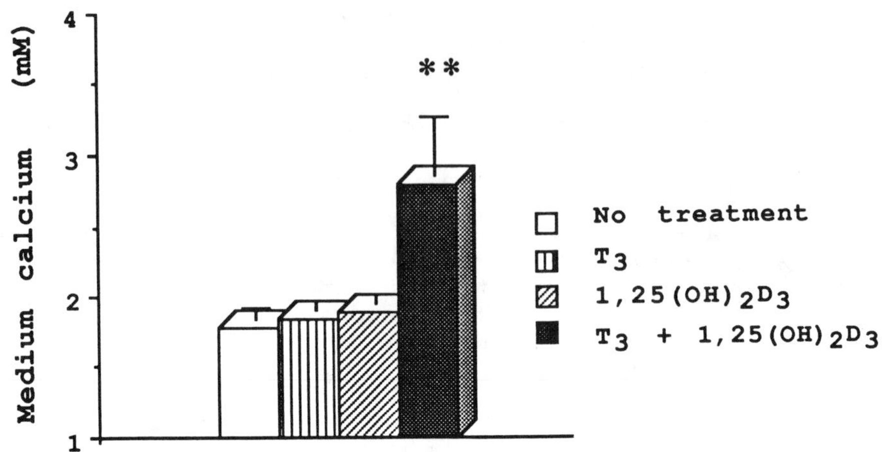

Figure 3: Effect of $1,25(OH)_2D_3$ and T_3 on resorption of cultured neonatal mouse calvaria. Bones were cultured as detailed in ref. 49. Hormone concentrations: T_3, $5x10^{-9}$ M; $1,25(OH)_2D_3$, $1x10^{-12}$ M. Medium calcium concentration was determined after 72 h culture time. Data are means ± SEM (n = 5) from a typical experiment. Significance of difference (analysis of variance):** $p < 00.1$ (vs. no treatment group)

ACKNOWLEDGEMENTS

Investigations of the authors referred to in this review were supported by grants from the Medizinisch-Wissenschaftlicher Fonds des Bürgermeisters der Bundeshauptstadt Wien, from the "Jubiläumsfonds der Österreichischen Nationalbank" (project No. 2922), and from the Anton Dreher Memorial Fund of the University of Vienna Medical School (grant nos. 73/86, 61/85 and 7/82).

The authors are greatly indebted to Drs. O. Hoffmann, E. Czerwenka, E. Lehner and N. Fratzl-Zelman, Ludwig Boltzmann Research Unit for Clinical and Experimental Osteology, Vienna, for performing experiments with organ-cultured neonatal mouse calvaria, as well as to Drs. H. Gleispach and H.-J.Leis, University of Graz, Austria, for PG determinations by gas chromatography/mass spectrometry.

REFERENCES

1. Perry III, H. M., Thyroid hormones and mineral metabolism, in *Bone and Mineral Research/6,* Peck, W. A., Ed., Elsevier, Amsterdam, 1989, 113
2. Adams, P. H., Jowsey, J., Kelly, P. J., Riggs, B. L., Kinney, V. R. and Jones J. D., Effect of hyperthyroidism on bone and mineral metabolism in man, *Q. J. Med.*, 141, 1, 1967
3. Auwerx J. and Bouillon, R., Mineral and bone metabolism in thyroid disease: a review, *Q. J. Med. New Series*, 60, 737, 1986
4. Moskilde, L. and Melsen, F., Morphometric and dynamic studies of bone changes in hyperthyroidism, *Acta Pathol. Microbiol. Scand. [A]*, 86, 56, 1978
5. Mundy, G. R. and Raisz, L. G., Thyrotoxicosis and calcium, *Metabolism* 2, 285, 1979
6. Mundy, G. R., Shapiro, J. L., Bandelin, J. G., Canalis, E.M. and Raisz, L. G., Direct stimulation of bone resorption by thyroid hormones, *J. Clin. Invest.*, 58, 529, 1976
7. Bouillon, R., Muls, E. and De Moor, P., Influence of thyroid function on the serum concentration of 1,25-dihydroxyvitamin D_3, *J. Clin. Endocrinol. Metab.*, 51, 793, 1980
8. Stumpf, W. E., Vitamin D - Soltriol. The heliogenic steroid hormone: somatotrophic activator and modulator. Discoveries from histochemical studies lead to new concepts, *Histochemistry,* 89, 209-219, 1988
9. Sar, M., Miller W. E. and Stumpf, W. E., Effects of 1,25(OH)$_2$vitamin D_3 on thyrotropin secretion in vitamin D deficient male rats, *Physiologist*, 24, 70, 1981
10. Törnquist, K. and Lamberg-Allardt, C. Effect of 1,25-dihydroxyvitamin D3 on TSH secretion from rat pituitary cells in culture. *Acta endocrinol. (Copenh.)*, 114, 357, 1987
11. Berg, J. P., Torjesen, P. A. and Haug. E., 1,25(OH)2D3 attenuates iodide uptake in a rat thyroid cell line, presented at *Eighth Workshop on Vitamin D,* Paris, July 5-10, 1991, 108
12. Lee, D.B.N., ed., Intestinal Absorption of Minerals: Experimental and Clinical, *Miner. Electrolyte Metab.*, 16, 93 1990
13. Bronner, F., Intestinal calcium transport: the cellular pathway, *Miner. Electrolyte Metab.*, 16, 94, 1990
14. Corradino, R. A. and Brewer, L. M., Hormonal regulation of tissue weight and responsiveness to vitamin D in the embryonic chick organ culture, in *Calcium and Phosphate Transport Across Biomembranes*, Bronner, F. and Peterlik, M., eds., Academic Press New York, 1981, 27
15. Cross, H. S., Pölzleitner, D. and Peterlik, M., Intestinal phosphate and calcium absorption: joint

regulation by thyroid hormones and 1,25-dihydroxyvitamin D3. *Acta endocrinol. (Copenh.)*, 113, 96, 1986

16. Cross, H. S. and Peterlik, M., Calcium and inorganic phosphate transport in embryonic chick intestine: triiodothyronine enhances the genomic action of 1,25-dihydroxycholecalciferol. *J. Nutr.*, 118, 1529, 1988
17. Cross, H. S. and Peterlik, M., Differentiation-dependent expression of 1,25(OH)2D3 actions on absorptive processes in cultured chick intestine: modulation by triiodothyronine, *Acta endocrinol. (Copenh.)*, 124, in press, 1991
18. Evans, R. M., The steroid and thyroid hormone receptor superfamily, *Science*, 240, 889, 1988
19. Schüele, R., Müller, M., Kaltschmid, C. and Renkawitz, R., Many transcription factors interact synergistically with steroid receptors, *Science*, 242, 1418, 1988
20. Perret, C., Bréhier, A., l`Horset, F., Gouhier, N. and Thomasset, M., Analysis of rat calbindin-9k promotor, regulation by 1,25(OH)2D3. *J. Bone Min. Res.*, 4(Suppl. 1), 700, 1989
21. Peterlik, M., Intestinal phosphate transport, in *The Enzymes of Biological Membranes*, vol. 3, Martonosi, A., Ed., Plenum Press, New York 1985, 287
22. Quamme, G. A. and Shapiro, R. J., Membrane controls of epithelial phosphate transport, *Can. J. Physiol. Pharmacol.*, 65, 275, 1987
23. Cross, H. S., Debiec, H. and Peterlik, M., Mechanism and regulation of intestinal phosphate absorption, *Miner. Electrolyte Metab.*, 16, 115, 1990
24. Fuchs, R., Graf, J. and Peterlik, M., Effects of 1α-dihydroxycholecalciferol on sodium-ion translocation across chick intestinal brush-border membrane, *Biochem. J.*, 230, 441, 1985
25. Cross, H. S., Debiec., H. and Peterlik, M., Thyroid hormone enhances the genomic action of 1,25(OH)2D3 in the small intestine, in *Molecular and Cellular Regulation of Calcium and Phosphate Metabolism*, Peterlik, M. and Bronner, F., Eds., Alan R. Liss, Inc., New York, 1990, 163
26. Debiec, H., Cross, H. S. and Peterlik, M., 1,25-Dihydroxycholecalciferol-related Na^+/D-glucose transport in brush-border membrane vesicles from embryonic chick jejunum: modulation by triiodothyronine, *Eur. J. Biochem.*, in press
27. Bronner, F., Renal calcium transport: mechanisms and regulation - an overview, *Am. J. Physiol.*, 257, F707, 1989
28. Noronha-Blob, L. and Sacktor, B., Inhibition by glucocorticoids of phosphate transport in primary cultured renal cells, *J. Biol. Chem.*, 261, 2164, 1986
29. Espinosa, R. E., Keller, M. J., Yusufi, A. N. K. and Dousa, T. P., Effect of thyroxine administration on

phosphate transport across renal cortical brush border membrane. *Am. J. Physiol.*, 246, F133, 1984

30. Stein, G. S., Lian, J. B. and Owen T. A., Relationship of cell growth to the regulation of tissue-specific gene expression during osteoblast differentiation, *FASEB J.*, 4, 3111, 1990
31. Morrison, N. A., Shine, J., Fragonas, J.-C., Verkest, M., McMenemy, M. L. and Eisman, J. A., 1,25-Dihydroxyvitamin D-responsive element and glucocorticoid repression in the osteocalcin gene, *Science*, 246, 1158, 1989
32. Rizzoli, R, Poser, J. and Bürgi, U., Nuclear thyroid hormone receptors in cultured bone cells, *Metabolism*, 35, 71, 1986
33. Mäenpää, P.H., Pirskanen A. and Mahonen A., Differential effects of dexamethasone, estradiol, triiodothyronine and retinoic acid on 1,25-dihydroxyvitamin D-stimulated osteocalcin synthesis in human osteosarcoma cells, *Calcif. Tissue Int.*, 48 (Suppl.), A22, 1991
34. Hoffmann, O., Klaushofer, K., Koller, K., Peterlik, M. Mavreas, T. and Stern, P. H., Indomethacin inhibits thrombin-, but not thyroxin-stimulated resorption of fetal rat limb bones, *Prostaglandins*, 31, 601, 1986
35. Klaushofer, K., Hoffmann, O., Gleispach, H., Leis, H.-J., Czerwenka, E., Koller, K. and Peterlik, M., Bone-resorbing activity of thyroid hormones is related to prostaglandin production in cultured neonatal mouse calvaria, *J. Bone Min. Res.*, 4, 305, 1989
36. Krieger, N. S., Stappenbeck, T. S. and Stern, P. H., Cardiotonic agent milrinone stimulates resorption in rodent bone organ culture, *J. Clin. Invest.*, 79, 444, 1987
37. Lakatos, P. and Stern, P. H., Thyroxine stimulated bone resorption is inhibited by cyclosporine A but not by TGF-beta$_1$, estradiol or stanozolol, *J. Bone Min. Res.*, 5 (Suppl. 2), S 129
38. Fratzl-Zelman, N., Klaushofer, K., Hoffmann, O, Lehner, E., Bentz, H., Koller, K. and Peterlik, M., A TGF-β-directed monoclonal antibody inhibits the effect of thyroid hormones on prostaglandin-mediated bone resorption in neonatal mouse calvaria, *Calcif. Tissue Int.*, 48 (Suppl.), A44, 1991
39. Klaushofer, K., Czerwenka, E., Koller, K., Leis, H.-J., Gleispach, H. and Peterlik, M., Calcitriol-stimulated bone resorption in neonatal mouse calvaria: effects of indomethacin and gamma interferon, *J. Bone Mineral Res.*, 2 (Suppl. 1), 455, 1987
40. Casey, M.L., Griffin, J. E., Korte, K., Mitchell, M. D. and MacDonald, P C., Response of human amnion cells in culture to 1,25-dihydroxycholecalciferol: Increased 25-hydroxycholecalciferol-24-hydroxylase activity and prostaglandin E_2 formation, *Am. J. Obst. Gynecol.* 155, 1272, 1986
41. Honda, A., Morita, I., Murota, S. and Mori, Y.,

Appearance of the arachidonic acid metabolic pathway in human promyelocytic leukemia (HL-60) cells during monocytic differentiation: Enhancement of thromboxane synthesis by 1,25-dihydroxyvitamin D_3, *Biochim. Biophys. Acta*, 877, 423, 1986

42. Rigby, W. F. C., Noelle, R. J., Krause, K. and Fanger, M. W., The effects of 1,25-dihydroxyvitamin D_3 on human T lymphocyte activation and proliferation: a cell cycle analysis, *J. Immunol.*, 135,2279, 1985

43. Koren, R., Ravid, A., Rotem, C., Shohami, E. Liberman, U. A. and Novogrodsky, A., 1,25-Dihydroxyvitamin D_3 enhances prostaglandin E_2 production by monocytes: a mechanism which partially accounts for the antiproliferative effect of $1,25(OH)_2D_3$ on lymphocytes, *FEBS Lett.*, 205, 113, 1986

44. Nijweide, P.J., Burger, L. H. and Feyen, J. H. M., Cells of bone: proliferation, and hormonal regulation. *Physiol. Rev.*, 66, 855, 1986

45. Rodan, S.B., Wesolowski, G. and Rodan, G. A., Clonal differences in prostaglandin synthesis among osteosarcoma cell lines. *J. Bone. Min. Res.*, 1, 213, 1986

46. MacDonald, B.R., Gallagher, J. A., Ahnfeld-Ronne, I., Beresford, J. N., Gowen, M. and Russel, R. G. G., Effects of bovine parathyroid hormone and 1,25-dihydroxyvitamin D_3 on the production of prostaglandins by cells derived from human bone, *FEBS Lett.*, 169, 49, 1984

47. Feyen, I.H., van der Wilt, G., Moonen, P., Di Bon, A., and Nijweide, P. J., Stimulation of arachidonic acid metabolism in primary cultures of osteoblast-like cells by hormones and drugs, *Prostaglandins*, 28, 769, 1984

48. Raisz, L.G. and Martin, T. J., Prostaglandins in bone and mineral metabolism, in *Bone and Mineral Research, Annual 2*, Peck, W. A., ed., Elsevier Science Publishers, Amsterdam 1983, 286

49. Stern, P.H. and Krieger, N. S., Comparison of fetal rat limb bones and neonatal mouse calvaria: Effects of parathyroid hormone and 1,25-dihydroxyvitamin D_3, *Calcif. Tissue Int.*, 35, 172, 1983

50. Lerner, U.H., Modifications of the mouse calvarial technique improve the responsiveness to stimulators of bone resorption. *J. Bone Min. Res.* 2, 375, 1987

51. Hoffmann, O., Klaushofer, K., Koller, K. and Peterlik, M., Prostaglandin-related bone resorption in cultured neonatal mouse calvaria: Evaluation of biopotency of nonsteroidal anti-inflammatory drugs, *Prostaglandins*, 30, 857, 1985

52. Boskey, A.L. and Wientroub, S., The effect of vitamin D deficiency on rat bone lipid composition. *Bone*, 7, 277, 1986

53. Schwartz, Z. and Boyan, B., The effects of vitamin D metabolites on phospholipase A_2 activity of growth zone and resting zone cartilage cells in vitro, *Endocrinology*, 122, 2191, 1988

54. Key, jr., L. L., Weichselbaum, R. R. and Carnes, jr., D. L., A link between calcitriol and bone resorption, *Bone and Mineral*, 3, 201, 1988
55. Hoffmann, O., Klaushofer, K., Gleispach, H., Leis, H.-J., Luger, T., Koller, K. and Peterlik, M., Gamma interferon inhibits basal and interleukin 1-induced prostaglandin production and bone resorption in neonatal mouse calvaria, *Biochem. Biophys. Res. Comm.*, 143, 38, 1987
56. Tashijan, A. H., Voelkel, E. F., Lazzaro, M., Singer, F. R., Roberts, A. B., Derynck, R., Winkler, M. E. and Levine, L., Alpha and beta human transforming growth factors stimulate prostaglandin production and bone resorption in cultured mouse calvaria. *Proc. Natl. Acad. Sci. USA* , 82, 4535, 1985
57. Tashijan, A. H. Jr., Voelkel, E. F., Lazzaro, M., Goat, D., Bosma, T. and Levine, L., Tumor necrosis factor α (cachectin) stimulates bone resorption in mouse calvaria via a prostaglandin mediated mechanism, *Endocrinology*, 120, 2029, 1987
58. Pfeilschifter, J. and Mundy, G. R., Modulation of type β transforming growth factor activity in bone cultures by osteotropic hormones, *Proc. Natl. Acad. Sci. USA*, 84, 2024, 1987
59. Bhalla, A.K., Amento, E. P. and Krane, S. M., Differential effects of 1,25-dihydroxyvitamin D_3 on human lymphocytes and monocyte/macrophages: inhibition of interleukin-2 and augmentation of interleukin-1 production, *Cell. Immunol.*, 98, 311, 1986

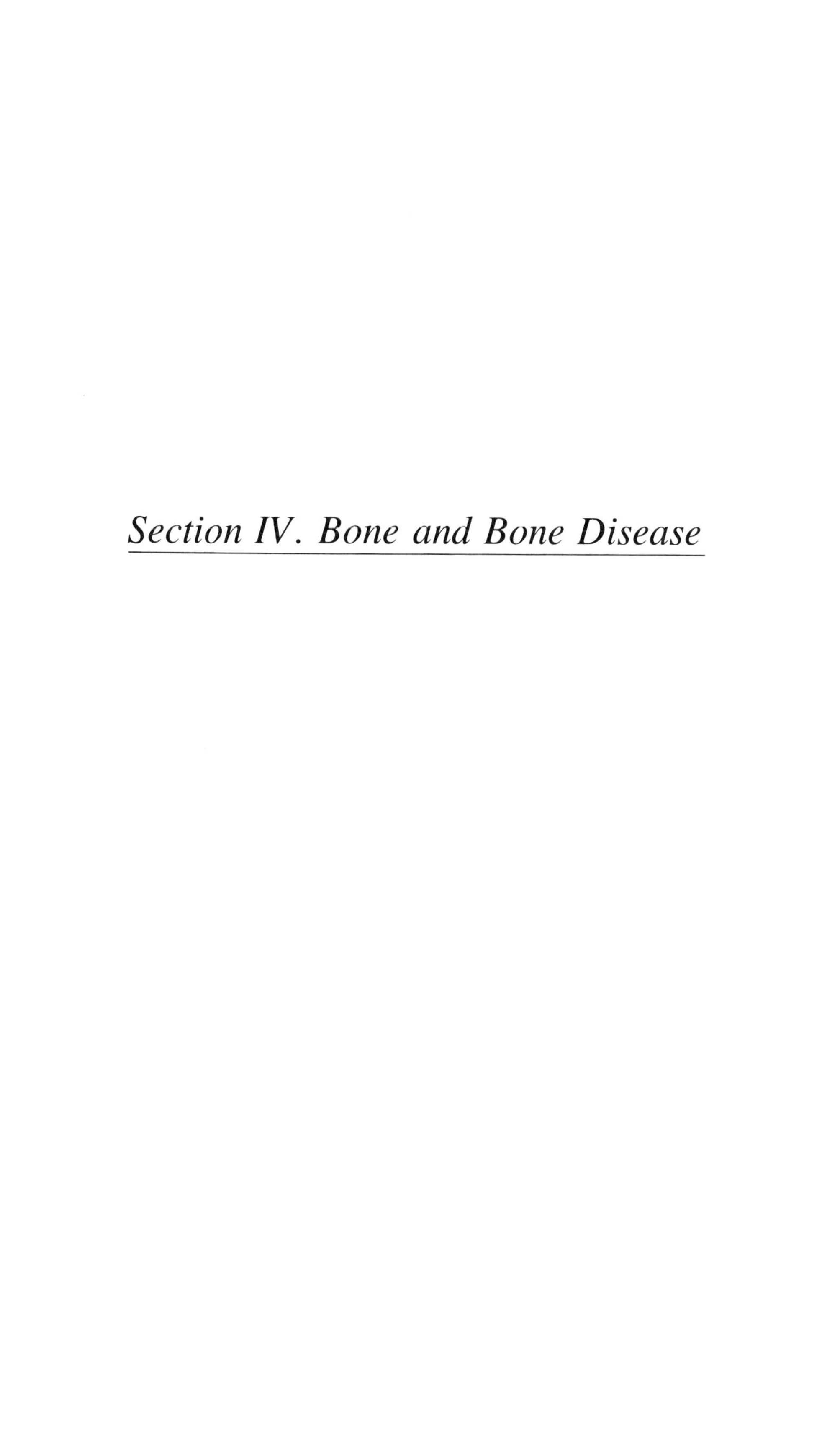

Section IV. Bone and Bone Disease

CONTROL OF CYTOSOLIC CALCIUM IN OSTEOCLASTS *IN VITRO*

ANNA TETI[1] AND ALBERTA ZAMBONIN ZALLONE[2]. INSTITUTE OF HUMAN ANATOMY, [1]SCHOOL OF PHARMACY, AND [2]SCHOOL OF MEDICINE, UNIVERSITY OF BARI, ITALY

The control of osteoclast activity is an intriguing problem, still poorly understood. The major activity of osteoclasts is the resorption of the calcified bone matrix which leads to the release of calcium and phosphate ions and of other digested bone matrix components in the extracellular microenvironment. Therefore, in the economy of the control of calcium homeostasis, osteoclast activity accounts for the increase of calcium concentration in the extracellular fluids.

The regulation of calcium homeostasis involves the activity of a number of different organs, including endocrine glands. Despite the clear *in vivo* overall effect of such regulation, the bone cellular targets for most committed factors remain unknown. Particularly, parathyroid hormone, 1,25-dihydroxyvitamin D_3, and a number of cytokines and local factors, which have been shown to influence bone resorption *in vivo*, fail to exert their effect on the resorbing cells *in vitro*, indicating that osteoclasts lack specific receptors or that these receptors are not functional[1-5].

Since osteoclast activity is mainly set to maintain calcium homeostasis, we explored the possibility that calcium ions might be directly involved in the control of osteoclast function. There are several examples of cellular activity directly regulated by the extracellular calcium concentration ($[Ca^{2+}]_o$) in a number of different species. The hormonal control of calcium homeostasis is mainly due to the activity of parathyroid chief cells and thyroid C cells, which secrete parathyroid hormone and calcitonin, respectively. Both hormones influence osteoclast activity. Calcitonin diminishes calcium concentration in the extracellular fluid by receptor-mediated inhibition of osteoclast function[6], which leads to arrest of motility and bone resorption[7,8]. Parathyroid hormone enhances the release of calcium from bone, but the cellular system affected is still uncertain[3,9]. The secretion of both hormones is directly regulated by the calcium concentration in the extracellular fluids[10,11]. This represents an important example of a cell activity directly influenced by calcium ions. Recent reports from our laboratory and from others clearly demonstrated that a similar cellular control regulates osteoclast activity. Osteoclast bone resorption is, in fact, directly inhibited by increased $[Ca^{2+}]_o$ in a dose-dependent manner. High $[Ca^{2+}]_o$ significantly reduces both the organic bone matrix degradation by avian osteoclasts[12] and pit formation by mammalian osteoclasts[13], indicating that a common mechanism of $[Ca^{2+}]_o$-induced osteoclast control is diffused in different species.

Bone resorption is not the only osteoclast cellular parameter influenced by high $[Ca^{2+}]_o$. In mammalian species cell retraction and inhibition of spreading have been observed[13,14], as well as a reduced rate of enzyme release. Important alteration of the cytoskeletal organization has been demonstrated in avian osteoclasts as well. Microfilaments represent important components of osteoclast cytoskeleton and their arrangement modifies according to the functional status of the cell[15]. They are particularly concentrated in the clear zone[16] of polarized osteoclasts, where they are assembled in the axis of membrane protrusions performing cell-to-substratum contacts with the bone surface[17-19]. These areas, named podosomes, represent peculiar focal adhesions[20] mediating, in a dynamic manner[21], the firm junction of the clear zone membrane with the bone surface to be resorbed. This provides the seal for the segregation of the resorbing compartment, where bone resorption takes place by the activity of the ruffled border membrane.

Interestingly, podosomes array a number of actin-binding, attachment and regulatory proteins[17,22], which modulate the assembly of microfilaments. These proteins represent specific targets for intracellular messengers such as $[Ca^{2+}]_i$ and protein kinases. Therefore, important regulative events occur at this level accounting for rearrangement of adhesion and change of cell polarization. Since polarization of osteoclasts permits the correct distribution of the molecular mechanisms for bone resorption, and adhesion to bone allows the segregation of the resorbing compartment where bone matrix degradation takes place, these two events are mandatory for osteoclast activity. We demonstrated that both events are affected, in avian osteoclasts, by $[Ca^{2+}]_o$ in a dose-dependent fashion. Increased $[Ca^{2+}]_o$ induces disassembly of podosomes with organization of microfilaments in an diffuse distribution, and causes osteoclasts to loose their polarization, as indicated by the disappearance of the clear zone[12]. These results clearly indicate that osteoclasts, like parathyroid chief cells and thyroid C cells, are directly regulated by $[Ca^{2+}]_o$, whose increase alters their morphology and inhibits the resorbing activity[15].

Several external signals affect cell functions through transduction mechanisms involving changes in the level of second messengers. These are intracellular molecules capable of inducing molecular modification of multiple cellular targets. $[Ca^{2+}]_o$ seems to alter osteoclast function by binding to a calcium-sensing receptor-like mechanism[23] sensitive to a number of different cations[13], which induces activation of phospholipase C. Phospholipase C is a membrane enzyme programmed for the hydrolysis of phosphoinosites in inositol-1,4,5 phosphate (IP_3) and diacylglycerol (DG). IP_3 is soluble, and induces receptor-mediated release of calcium from the intracellular stores. DG remains in the plasma membrane and activates protein kinases C (PKC), a class of enzymes that, by substrate phosphorylation, changes cell metabolism. We found that, following receptor binding, high $[Ca^{2+}]_o$ affects this signal transduction mechanism inducing increase of $[Ca^{2+}]_i$[13]. Such increase is due to a complex phenomenon since it depends by release of calcium from the cellular stores and by gating of calcium-operated, dihydropiridine insensitive, calcium channels. These are not the only calcium channels in osteoclasts, since L-type, dihydropiridine sensitive voltage operated calcium channels have also been found[12]. Similar to high $[Ca^{2+}]_o$, also the gating of these voltage-dependent calcium channels leads to increase of $[Ca^{2+}]_i$ and has an inhibitory role over the osteoclast function.

We also demonstrated that PKC, which possibly is activated in osteoclasts by increase of $[Ca^{2+}]_o$, alters osteoclast function inducing modification of cytoskeletal arrangement and bone resorption. The role played by PKC seems to be consistent with inhibition of the cell activity, and with a positive feed-back control of the membrane calcium sensing mechanism. In fact, activation of the enzyme, potentiates the amplitude of $[Ca^{2+}]_o$-induced $[Ca^{2+}]_i$ response[24]. Similar positive potentiation has been demonstrated to be exerted by cAMP[13] which represents the second messenger generated by calcitonin[6]. This inhibiting hormone, in fact, mimics the cAMP-induced potentiation of the calcium sensing-dependent response, indicating a synergistic role played by the two signals, calcitonin and $[Ca^{2+}]_o$, on the osteoclast activity[13]. Since in the intact animal calcitonin is secreted by the thyroid C cells when serum calcium level is high, this synergism may have an important regulatory role also *in vivo*.

These results indicate that an important osteoclast regulating pathway has been identified, which accounts for a direct modulation of the cell morphology and resorbing activity. This pathway is similar to that described in the parathyroid chief cells, in which high $[Ca^{2+}]_o$ inhibits cell activity and induces increase of $[Ca^{2+}]_i$[25].

Modification of $[Ca^{2+}]_i$ in osteoclasts have been demonstrated to be involved also in osteoclast activation. In fact, when $[Ca^{2+}]_i$ is reduced, for example by metabolic acidosis, cell polarization and adhesion[26], and bone resorption[27] are enhanced. Similar modification of the $[Ca^{2+}]_i$ have been observed when osteoclasts interact with a number of extracellular proteins, including the bone proteins osteopontin and bone sialoprotein II, via a receptor similar to the $\alpha_v\beta_3$ integrin[28]. This indicates that osteoclast activation may simply occurs by contact with the bone matrix, which per se could represent the mandatory exogenous stimulus for bone resorption. Following this indication, we can assume that differentiated osteoclasts are programmed to resorb bone, and that initiation of bone resorption may depend upon the availability of a lining cell-free bone surface to be contacted, which induces both adhesion and polarization. These two events prime the molecular mechanism of bone resorption. Inhibiting stimuli, such as calcitonin or increased $[Ca^{2+}]_o$, are therefore necessary for blocking osteoclast activity. Other exogenous stimuli, such as retinol[29], or metabolic acidosis[26], may enhance the basal resorbing activity, modulating osteoclast function.

In conclusion, new aspects of osteoclast control have recently been described. In this regard, calcium ions play crucial roles either as exogenous signal, either as intracellular messenger. New perspective may, therefore be hypothesized, particularly for therapeutic treatment of severe metabolic bone diseases.

REFERENCES

1. Silve, C.M., Hradek, G.T., Jones A.L., Arnaud, C.D., Parathyroid Hormone receptor in intact embryonic chicken bone: characterization and cellular localization, J. Cell Biol. 94, 379, 1982.
2. Rao, L.G., Murray, T.M., Heersche, J.N.M., Immunohistochemical demonstration of parathyroid hormone binding to specific cell types in fixed rat bone tissue, Endocrinology, 113, 805, 1983.
3. Teti, A., Rizzoli, R., Zambonin Zallone, A., Parathyroid hormone binding to cultured avian osteoclasts, Biochem. Biophys. Res. Comm., 174, 1217, 1991.
4. Merke, J., Klaus, G., Hugel, U., Waldherr, R., Ritz, E., No 1,25-dihydroxyvitamin D_3 receptors on osteoclasts of calcium-deficient chicken despite demonstrable receptors on circulating monocytes, J. Clin. Invest., 77, 312, 1986.
5. Thomson, B.M., Saklatvala, J., Chambers, T.J., Osteoblasts mediate interleukin 1 stimulation of bone resorption by rat osteoclasts, J. Exp. Med., 164, 104, 1986.
6. Nicholson, G.C., Moseley, J.M., Sexton, P.M., Mendelsohn, F.A.O., Martin, T.J., Abundant calcitonin receptors in isolated rat osteoclasts, J. Clin. Invest., 78, 355, 1986.
7. Chambers, T.J., Chambers J.C., Symonds, J., Darby, J.A., The effect of human calcitonin on the cytoplasmic spreading of rat osteoclasts, J. Clin. Endocrinol, Metab., 63, 1080, 1986.
8. Chambers, T.J., McSheehy P.M.J., Thomson, B.M., Fuller, K, The effect of calcium-regulating hormones and prostaglandins on bone resorption by osteoclasts disaggregated from neonatal rabbit bones, Endocrinology, 116, 234, 1985.

9. Rodan, G.A., Martin, T.J., Role of osteoblasts in hormonal control of bone resorption, A hypothesis, Calcif. Tissue Int., 33, 349, 1981.
10. Shoback, D., Tharcher, J., Leombruno, R., Brown, E.M., Effects of extracellular Ca^{++} and Mg^{++} on cytosolic Ca^{++} and PTH release in dispersed bovine parathyroid cells, Endocrinology, 113, 424, 1983.
11. Fried, R.M., Tashjian, A.M., Jr., Unusual sensitivity of cytosolic free Ca^{++} to changes in extracellular Ca^{++} in rat C cells, J. Biol. Chem., 261, 7669, 1986.
12. Miyauchi, A., Hruska, K.A., Greenfield, E.M., Randall, D., Alvarez, J., Barattolo, R., Colucci, S., Zambonin Zallone, A., Teitelbaum, S.L., Teti, A., Osteoclast cytosolic calcium, regulated by voltage operated calcium channels and extracellular calcium, controls podosome assembly and bone resorption, J. Cell Biol., 111, 2543, 1990.
13. Malgaroli, A., Meldolesi, J., Zambonin Zallone, A., Teti, A., Control of cytosolic free calcium in rat and chicken osteoclasts. The role of extracellular calcium and calcitonin, J. Biol. Chem., 264, 14342, 1989.
14. Zaidi, M., Datta H.K., Patchell, A., Moonga, B., MacIntyre, I., Calcium-activated intracellular calcium elevation: a novel mechanism of osteoclast regulation. Biochem. Biophys. Res. Comm., 163, 1461, 1989.
15. Teti, A., Marchisio, P.C., Zambonin Zallone A., The clear zone in the osteoclast function. The role of podosomes in the regulation of the bone resorbing activity, Am. J. Physiol., 1991, in press.
16. King, GJ., Holtrop, M.E., Actin-filaments in bone cells of cultured mouse calvaria as demonstrated by binding to heavy meromyosin, J. Cell Biol., 66, 445, 1975.
17. Marchisio, P.C., Naldini, L., Cirillo, D., Primavera, M.V., Teti, A., Zambonin Zallone, A., Cell-substratum interaction of cultured avian osteoclasts is mediated by specific adhesion structures, J. Cell Biol., 99, 1696, 1984.
18. Zambonin Zallone, A., Teti, A., Carano, A., Marchisio, P.C., The distribution of podosomes in osteoclasts cultured on bone laminae: effect of retinol, J. Bone Min. Res., 3, 517, 1988.
19. Turksen, K., Kanehisa, J., Opas, M., Heersche, J.N.M., Adhesion patterns and cytoskeleton of rabbit osteoclasts on bone slices and glass, J. Bone Min. Res., 3, 389, 1988.
20. Burridge, K., Fath, K., Kelly, T., Nuckolls, G., Turner, C., Focal adhesions: transmembrane junctions between the extracellular matrix and the cytoskeleton. Ann. Rev. Cell Biol., 4, 487, 1988.
21. Kanehisa, J., Yamanaka, T., Doi, S., Turksen, K., Heersche, J.N.M., A band of F-actin containing podosomes is involved in bone resorption by osteoclasts, Bone 11, 287, 1990.

22. Marchisio, P.C., Cirillo, D., Teti, A., Zambonin Zallone, A., Tarone, G., Rous sarcoma virus transformed fibroblasts and cells of monocytic origin display a peculiar dot-like organization of cytoskeletal proteins involved in microfilament-membrane interactions, Exp. Cell Res., 169, 202, 1987.
23. Miyauchi, A., Greenfield, E.M., Alvarez, J., Bar-Shavit, Z., Huskey, M., Teitelbaum, S., Hruska, K., The osteoclast calcium/divalent cation receptor stimulates subcellular Ca^{2+} gradients and phospholipase C activation, J. Bone Min. Res., suppl. vol. 5, 299, 1990.
24. Teti, A., Colucci, S., Grano, M., Argentino, L., Zambonin Zallone, A., Regulation of osteoclast activity: protein kinase C affects cytoskeleton organization and bone resorption, and regulates extracellular calcium-sensing in isolated osteoclasts in vitro, submitted for publication.
25. Nemeth, E.F., Scarpa, A., Rapid mobilization of cellular Ca^{++} in bovine parathyroid cells evoked by extracellular divalent cations.
26. Teti, A., Blair, H.C., Schlesinger, P., Grano, M., Zambonin Zallone, A., Kahn, A.J., Teitelbaum, S.L., Hruska, K.A., Extracellular protons acidify osteoclasts, reduce cytosolic calcium and promote expression of cell-matrix attachment structures, J. Clin. Invest., 84, 773, 1989.
27. Arnett, T.R., Dempster D.W., Effect of pH on bone resorption by rat osteoclasts in vitro, Endocrinology, 119, 119, 1986.
28. Teti, A., Ross, P., Miyauchi, A., Teitelbaum, S., Hruska, K., Grano, M., Colucci, S., Gehron-Robey, P., Zambonin Zallone, A., Osteopontin and BSP II trigger changes in cytosolic free calcium concentration in chicken osteoclasts. Calcif. Tissue Int., suppl. vol. 48, 21, 1991.
29. Oreffo, R.O.C., Teti, A., Francis, M.J.O., Triffitt, J.T., Carano, A., Zambonin Zallone, A., Effect of vitamin A on bone resorption: evidence of direct stimulation of isolated chicken osteoclasts by retinol and retinoic acid. J. Bone Min. Res., 3, 203, 1988.

SIDE EFFECTS OF DRUGS ON BONE

Paula H. Stern, Ph.D.
Department of Pharmacology, Northwestern University,
Chicago, IL 60611, USA

Pharmacologic selectivity is a highly desirable, but rarely attainable goal. Although the various organs and tissues of the body carry out unique functions, many of the same signalling processes, membrane transport systems and metabolic pathways are utilized. Unless a drug can be targeted to a tissue-specific receptor, enzyme, or other constituent unique to that tissue, and there are no further systemic consequences of that interaction, side effects are inevitable. The intensity of the side effects will depend upon the extent of overlap of the functions, and the sensitivity and importance of these functions in the non-targetted tissue. Side effects are not necessarily deleterious. If they are beneficial, they may lead to the development of new therapies directed at the secondary tissue.

The focus of this discussion on side effects of drugs on bone will be on drugs with side effects that are deleterious, e.g., those causing osteoporotic or osteomalacic conditions. The discussion also includes some drugs for which there was a logical expectation of side effects that have not materialized. The occurrence of side effects has provided insights into the biology of the mineralized tissues and the relative importance of certain processes, cells and mediators. Conversely, an improved understanding of the basic biology should thus lead to better prediction of side effects and facilitate the design of new and better primary therapeutic agents for the treatment of bone disease.

Table 1 lists a number of drugs that have side effects on bone and/or calcium metabolism when used therapeutically for other indications. The left-hand column are hormonal agents, although some may be used extensively for non-hormonal applications, as well. Several other drugs with side effects on bone are listed in the right-hand column. This discussion will not be comprehensive. Rather, it will focus on several of the newer agents, including immunosuppressants, antiestrogens, and hypothalamic gonadotropin-releasing hormones. Glucocorticoids and thyroid hormones cannot be ignored in any discussion of the effects of drugs on bone. They are also addressed elsewhere in this symposium as etiologic factors in secondary osteoporosis.

TABLE 1.
Drugs with Side Effects on Bone and Calcium Metabolism

Glucocorticoids	Anticonvulsants
Thyroxin	Eicosanoids, Antiinflammatory agents
Gonadal hormones	Immunosuppressants
GnRH	Thiazide diuretics
Antiestrogens	Retinoids
	Heparin

I. GLUCOCORTICOIDS

Adrenal cortical steroids have multiple effects on bone (Table 2). Long term exposure to elevated concentrations of glucocorticoids leads to osteopenia. A high incidence of osteoporosis in patients with Cushing's syndrome was documented about 40 years ago.[1] Since glucocorticoids are widely used for a number of indications, including inflammatory connective tissue diseases such as rheumatoid arthritis and allergic states such as asthma, the likelihood of these side effects is significant. Data on the incidence and prevalence of side effects on bone with different therapeutic steroid regimens are limited. The picture is particularly complicated in disorders such as rheumatoid arthritis in which decreased physical activity may predispose patients to osteoporosis. One study revealed that women with rheumatoid arthritis treated with 8.5 mg/day prednisolone had lumbar spine bone mineral densities that were not lower than those in women with rheumatoid arthritis who had not received steroids.[2] Comparison of biochemical parameters of bone metabolism in patients with rheumatoid arthritis receiving gold salts or penicillamine and patients treated with glucocorticoids revealed that fasting urinary calcium/creatinine and hydroxyproline/creatinine ratios were increased in patients treated with 2.5-10 mg/day of prednisone.[3] Both studies suggested differences in susceptibility based upon sex and dose of steroid.

Mechanisms of increased bone loss due to steroids are likely to be multifactorial. Decreased intestinal calcium reabsorption and renal tubular calcium reabsorption may result in secondary hyperparathyroidism in glucocorticoid-treated patients.[4] Nephrogenous cyclic AMP is elevated in patients receiving glucocorticoid therapy.[5] At the cellular level in bone, glucocorticoids increase the cyclic AMP responses to parathyroid hormone[6] and can alter the density of receptors for calcitriol, possibly by inhibiting their degradation.[7] Antianabolic effects of glucocorticoids on bone include the inhibition of collagen synthesis, probably at the transcriptional level.[8] Glucocorticoids have been shown to inhibit the replication of periosteal cells.[9] Secondary effects of glucocorticoids on the secretion and action of local factors in bone may also play a role in the observed effects. Glucocorticoids influence prostaglandin synthesis by inhibiting phospholipase A_2.[10] Prostaglandins have effects leading to both enhanced mineralization and increased resorption.[11] Also, glucocorticoids have effects on both the production and action of various lymphokines;[12] many of these, including IL-1, TNF and IFN-γ are active on bone.

Glucocorticoids are used therapeutically on a short term basis to treat hypercalcemia. The direct effects of glucocorticoids on bone resorption, in studies of several days' duration, are to inhibit hormonally-stimulated bone loss.[13] Thus, direct effects of glucocorticoids on resorption seem more likely to play a significant role in the antihypercalcemic effects than in the osteopenic responses. The action of glucocorticoids to attenuate the hypercalcemic or bone-resorbing effects is characterized by selectivity; effects of certain agents, e.g. calcitriol, interleukins, are more readily reversed or prevented that those of others, e.g. parathyroid hormone, thyroid hormones.[13] This selectivity suggests that the pathways or local co-factors may differ and that glucocorticoids may act by selectively inhibiting the production of cytokines that act as co-factors.

TABLE 2.
Effects of Glucocorticoids on Bone

A. Clinical observations
   1) Osteopenia
      Related to dose, duration, osteoporosis risk factors
      Biochemical parameters often normal

   2) R_x of hypercalcemia: Etiology determines antihypercalcemic efficacy

B. In vitro effect: Inhibition of resorption

   Question: What are the relative contributions of:

   1) Secondary hyperparathyroidism
   2) Antianabolic effects
   3) Receptor regulation
   4) Actions on autocrine/paracrine synthesis and secretion

   to the overall effects of glucocorticoids on bone?

II. THYROID HORMONES

Bone remodelling is increased in hyperthyroidism and even patients successfully treated for hyperthyroidism may have increased risk for osteoporosis due to decreased bone mass (Table 2).[15,16] Individuals receiving exogenous thyroid hormones as replacement or suppressive therapy might therefore also be at risk for elevated rates of bone loss. This has been documented in a number of studies. Decreased bone density in the spine and femur has been found in women receiving long-term L-thyroxin therapy, e.g. L-thyroxin for a minimum of five years.[17-20] Biochemical measurements are consistent with a stimulation of bone turnover, with elevated alkaline phosphatase, osteocalcin and urinary hydroxyproline.[21]

Thyroid hormone treatment of bones *in vitro* results in enhanced resorption.[22-24] Collagen degradation is stimulated by triiodothyronine in bone cultures[25] and alkaline phosphatase is increased in bone cultures and osteoblastic cells.[26,27] Findings in *in vitro* organ culture studies suggest complexities in the effects of thyroid hormones on bone resorption. Complete resorption is not usually observed. Biphasic effects have been reported. In one study [22], 10 μM was less effective than 1 μM in stimulating resorption. In another study, 0.01-0.3 μM concentrations resulted in elevations in bone calcium, phosphate and hydroxyproline with diminished release of incorporated ^{45}Ca, whereas 1 μM and 10 μM concentrations elicited loss of bone[28]. Both nuclear receptors[27,29] and non-genomic second messengers[30] are associated with responses of bone tissue to thyroid hormones. In calvaria but not in limb bones, indomethacin treatment prevents the bone-resorbing effects of thyroid hormones.[23,24]

TABLE 3.
Thyroid Hormone Effects on Bone

A. Clinical observations
Hypercalcemia in thyrotoxicosis
Decreased bone mass in patients on thyroid replacement
Biochemical indices of increased bone turnover

B. *In vitro* effects
Increased bone resorption
Qualitatively different from PTH (partial, biphasic ?)
Genomic and membrane mechanisms?

III. AGENTS AFFECTING GONADAL HORMONE SECRETION AND/OR ACTION

A. GONADOTROPIN RELEASING HORMONE (GnRH) (Table 4)

Continuous treatment with GnRH over a period of several weeks results in decreased secretion of gonadal hormones due to the desensitization of receptors on pituicytes. There are several therapeutic applications for continuous GnRH therapy, including palliative therapy for prostatic cancer, treatment of precocious puberty and treatment of endometriosis. Reductions in trabecular bone mass were noted in women receiving 200 - 1200 μg/day (by different routes) for 6 months or more.[31-32] *In vitro* experiments failed to show any direct effects of GnRH or the long-acting GnRH analog buserelin on control or parathyroid hormone-stimulated bones.[33] The bone loss observed in the therapeutic regimens is likely a consequence of diminished estrogen secretion. The challenge will be to design dosage regimens for GnRH that achieve the therapeutic objectives and preserve bone.

TABLE 4.
Effects of GnRH on Bone

A. Clinical observation: Decreased bone mass, increased fracture risk in women receiving GnRH for endometriosis

B. *In vitro*: No direct effects on bone resorption

Probable mechanism: Decreased estrogen

B. ANTIESTROGENS (Table 5) - The triphenylethylene compounds that have antiestrogenic activity appear to function by a variety of mechanisms, including weak agonism at estrogen receptors, cytotoxic actions mediated through specific antiestrogen receptors, and inhibitory actions on calmodulin and protein kinase C.[34,35] Estrogens affect prostaglandin synthesis, and tamoxifen has been shown to inhibit prostaglandin snthetase.[35,36] In view of these multiple mechanisms it is not surprising that the species, the tissue, the response being studied and even the conditions can determine the nature of the effect observed. Tamoxifen is a useful drug in the treatment of breast cancer, both in initial stages and as adjuvant therapy in advanced carcinoma. Studies of the bone mineral content in women with breast cancer who have been treated with tamoxifen have resulted in a range of findings. In one study in which bone density was assessed at 6 and 12 months after treatment with oral tamoxifen, 10 mg twice daily, there was a significant difference after 12 months between the patients who received tamoxifen and normal controls.[37] However, other studies of women receiving tamoxifen for breast cancer or mastalgia failed to demonstrate any increase in bone mass.[38,39]

Animal studies have also shown a variety of responses. Tamoxifen, 0.4 mg/day decreased bone mass in intact female rats.[40] In contrast, clomiphene and tamoxifen prevented bone loss in ovariectomized rats.[41,42] Similar findings were observed in the bone loss caused by tenotomy,[43] However, tamoxifen did not affect the bone loss elicited by calcitriol or by hypocalcemia.[44] In _in vitro_ experiments antiestrogens have produced pronounced inhibitory effects on bone. In organ culture the antiestrogens cause irreversible inhibition of resorption elicited by several stimulators[45]. The relationship between the direct effects, the observations in experimental animals and the clinical findings is unclear. However, it is likely that several of the different actions of the antiestrogens are involved. As newer, more selective antiestrogens are introduced, the possibility that they may elicit greater bone loss is cause for concern.

TABLE 5.
Effects of Antiestrogens on Bone

A. Clinical observations:
  Bone mass increased, unchanged
  (20 - 40 mg/day; 6 mos - 5 yr)

B. Animal studies:
  1) Bone loss from antiestrogen
  2) Prevention of bone loss: ovariectomy, tenotomy
  3) Failure to prevent bone loss: $1,25\text{-}(OH)_2D_3$, hypocalcemia

C. _In vitro_ studies:
  Inhibition of bone resorption (irreversible, insurmountable)

Mechanism(s): Estrogenic, antiestrogenic, other?

IV. IMMUNOSUPPRESSANTS

Immunosuppressant drugs, predominantly cyclosporine, have been widely used to prevent rejection after organ transplantation. They have also been investigated and had more limited application in the treatment of disorders with an early immune etiology, including diabetes mellitus and autoimmune disorders such as rheumatoid arthritis. It is only recently that reports of side effects on bone are beginning to emerge from these clinical studies (Table 6). Decreased bone mass, associated with high turnover bone remodelling has been noted in patients receiving cyclosporine after cardiac transplantation.[46] On the other hand, there might be beneficial effects on bone loss in patients with rheumatoid arthritis, where cyclosporine treatment results in decreased inflammation.[47]

Disparity between bone effects of cyclosporine in the presence and absence of inflammation is seen animal experiments. Bone loss has been observed in normal rats treated with 15 mg cyclosporine/kg/day for 28 days.[48] These rats also had elevated concentrations of circulating calcitriol. In contrast, the adjuvant arthritis model AIA rat, which has low calcitriol, showed increased bone density after treatment with 5-15 mg/kg/day for 10 days.[49] These results suggest a role for elevated circulating calcitriol in the cyclosporine-induced bone loss. However, not all findings are consistent with such a mechanism.[49] *In vitro*, cyclosporine and other immunosuppressive agents have invariably been inhibitory.[50-52] The inhibition is quite nonspecific, suggesting a toxic effect. However, it also was quickly reversible, readily surmountable, and generally not associated with inhibited macromolecular synthesis, factors inconsistent with a toxic effect. It is possible that an initial inhibitory effect could result in the activation of processes that stimulate bone turnover, since in at least one clinical study a decrease in serum calcium preceded the stimulatory changes.[53] The effects of the immunosuppressive agents on cytokine production may well play a major role in the effects of these agents on bone. This could explain the protective effects in inflammatory states, where the overproduction of inflammatory and osteolytic mediators is reduced by cyclosporine. The newer, more potent immunosuppressive agents are likely to also have effects on bone.

TABLE 6
Effects of Immunosuppressants on Bone

A. Clinical observations
   1. High turnover bone remodelling in transplantation patients
   2. Potential for decreased bone loss in inflammatory disease

B. Animal studies
   1. 'High turnover' bone loss in normal rats
   2. Increased bone density in AIA rat
      R_x: 5-15 mg/kg/day for 10 days (low 1,25-$(OH)_2D_3$)

C. *In vitro*
   Resorption inhibited by CsA and other immunosuppressants

V. ANTICONVULSANTS

Anticonvulsant treatment can result in osteopenia (Table 7). There are a number of contributory risk factors, including the therapeutic regimen, environment and concomitant therapy. Characteristics of both osteomalacia and osteoporosis have been found to occur. Actions of anticonvulsants that could contribute to the pathologic effects include inhibition of vitamin D activation, direct inhibition of intestinal calcium transport, and inhibition of bone resorption and collagen synthesis.[54] A recent report of elevated γ-carboxyglutamic acid-containing protein in children on anticonvulsant therapy is consistent with a state of high bone turnover.[55] This very complex and intriguing set of side effects on bone merits more extensive investigation.

TABLE 7.
Effect of Anticonvulsants on Bone

A. Clinical observations
osteomalacia, decreased bone mass, fractures, hypocalcemia, normal or decreased 25-OH-D, increased PTH, alkaline phosphatase

B. Determinants:
dose, duration, vitamin D, sunlight, physical activity, therapy affecting phosphate excretion, pH, hepatic P-450 enzymes

C. Mechanisms: Could involve actions on:
Vitamin D metabolism
Intestinal calcium transport
Bone resorption
Collagen synthesis

REFERENCES

1. Plotz, C.M., Knowlton, A.I., and Ragan, C. The natural history of Cushing's syndrome. Amer. J. Med. 13,597,1952.
2. Sambrook, P.N., EIsman, J.A., Champion, G.D., Yeates, M.G., Popcock, N.A. and Eberl, S. Determinants of axial bone loss in rheumatoid arthritis. Arthritis and Rheumatism 30,721,1987.
3. Als, O.S., Riis, B.J., Gotfredsen, A., Christiansen, C. and Deftos, L.J. Biochemical markers of bone turnover in rheumatoid arthritis. Acta. Med. Scand. 219,209,1986.
4. Fucick, R.F., Kukreja, S.C., Hargis, G.K., Bowser, E.N., Henderson, W.J. and Williams, G.A. J. Clin. Endoc. Metab. 49,152,1975.
5. Suzuki, Y., Ichikawa, Y., Saito, E., Homma, M. Importance of increased urinary calcium excretion in the development of secondary hyperparathyroidism of patients under glucocorticoid therapy. Metabolism 32,151,1983.
6. Chen, T.L. and Feldman, D. Glucocorticoid potentiation of the adenosine 3'5'-monophosphate response to parathyroid hormone in cultured rat bone cells. Endocrinology 102,589,1978.
7. Manolagas, S.C., Anderson, D.C. and Lumb, G.A. Glucocorticoids regulate the concentration of 1,25-$(OH)_2$-vitamin D_3 receptors in bone. Nature 277,314,1979.
8. Oikarinen, J. and Ryhanen, L. Cortisol decreases the concentration of translatable type I procollagen mRNA species in the developing chick embryo calvaria. Biochem. J. 198,519,1981.
9. Chyun, Y.S. and Raisz, L.G. Stimulation of bone formation by prostaglandin E_2. Prostaglandins 27,97,1984.
10. Flower, R.J. and Blackwell, G.J. Anti-inflammatory steroids induce biosynthesis of a phospholipase A_2 inhibitor which prevents prostaglandin generation. Nature 278,456,1979.

11. Raisz, L.G. and Martin, T.J. Prostaglandins in bone and mineral metabolism, in Bone and Mineral Research: Annual 2, Peck, W.A., ed. Elsevier, Amsterdam, 1984,286.
12. Dinarello, R.B. and Mier, J.W. Lymphokines. New Eng. J. Med. 317,940,1987.
13. Stern, P.H. Inhibition by steroids of parathyroid hormone-induced Ca^{45} release from embryonic bone in vitro. J. Pharmacol. 168,211,1969.
14. Raisz, L.G., Luben, R.A., Mundy, G.R., Dietrich, J.W., Horton, J.G. and Trummel, C.L. Effect of osteoclast activating factor from human lymphocytes on bone metabolism. J. Clin. Invest. 56,408,1975.
15. Melsen, F. and Mosekilde, L. Morphometric and dynamic studies of bone changes in hyperthyroidism. Acta Pathol. Microbiol. Scand. 85,141,1977.
16. Mosekilde, L., Kricksen, E.F. and Charles, P. Effects of thyroid hormones on bone and mineral metabolism. Metabolic Bone Disease, Part II pg.35.
17. Fallon, M.D., Perry, H.M.,II, Bergfeld, M., Droke, D., Teitelbaum, S.L. and Avioli, L.V. Arch. Int. Med. 143,442,1983.
18. Ross, D.S., Neer, R.M., Ridgway, E.C. and Daniels, G.H. Subclinical hyperthyroidism and reduced bone density as a possible result of prolonged suppression of the pituitary-thyroid axis with L-thyroxin. Amer. J. Med. 82,1167,1987.
19. Paul, T.L., Kerrigan, J., Kelly, A.M., Braverman, L.E. and Baran, D.T. Long-term L-thyroxine therapy is associated with decreased hip bone density in premenopausal women. J. Amer. Med. Assn. 259,3137,1988.
20. Stall, G.M. Harris, S., Sokoll, L.J. and Dawson-Hughes, B. Accelerated bone loss in hypothyroid patients overtreated with L-thyroxine. Ann. Int. Med. 113, 265, 1990.
21. Martinez,M.E., Herranz, L., de Pedro, C. and Pallardo, L.F. Osteocalcin levels in patients with hyper- and hypothyroidism. Horm. Metab. Res. 18,212,1986.
22. Mundy, G.R., Shapiro, J.L., Bandelin, J.G., Canalis, E.M. and Raisz, L.G. Direct stimulation of bone resorption by thyroid hormones. J. Clin. Invest. 58,529,1976.
23. Hoffmann, O., Klaushofer, K., Koller, K., and Peterlik, M. Evidence for a role of prostaglandins in thyroid hormone-induced bone resorption. J. Bone Min. Res. 1,102,1986.
24. Hoffmann, O. Klaushofer, K., Koller, K., Peterlik, M., Mavreas, T. and Stern, P. Indomethacin inhbits thrombin-, but not thyroxin-stimulated resorption of fetal rat limb bones. Prostaglandins 31,601,1986.
25. Halme, J. Vitto, J., Kivirikko, K.I., Saxen, L. Effect of triiodothyronone on the metabolism of collagen in cultured embryonic bones. Endocrinology 90,1476,1972.
26. Stracke, H. Rossol, S. and Schatz, H. Alkaline phosphatase and insulin-like growth factor in fetal rat bone under the influence of thyroid hormones. Horm. Metab. Res. 18,794,1986.
27. Sato, K., Han,D.C., Fuji, Y. Tsushima, T. and Shizume, K. Thyroid hormone stimulates alkaline phosphatase activity in cultured rat ostoeblastic cells (ROS 17/2.8) through 3,5,3'-triiodo-L-thyronine nuclear receptors. Endocrinology 120,1873,1987.

28. Soskolne, W.A., Schwartz, Z., Goldstein, M. and Ornoy, A. The biphasic effect of triiodothyronine compared to bone resorbing effect of PTH on bone modelling of mouse long bone in vitro. Bone 11, 301,1990.
29. Krieger, N.S., Stappenbeck, T.S. and Stern, P.H. Characterization of specific thyroid hormone receptors in bone. J. Bone Min. Res. 3,473,1988.
30. Lakatos, P. and Stern, P.H. Triiodothyronine stimulates the inositol phosphate second messenger pathway in bone. XXIInd European Symposium on Calcified Tissues, 1991.
31. Matta, W.H., Shaw, R.W., Hesp, R. and Evans, R. Reversible trabecular bone density loss following induced hypo-oestrogenism with the GnRH analog buserelin in premenopausal women. Clin. Endocrinol. 29,45,1988.
32. Dawood, M.Y., Lewis, V. and Ramos, J. Cortical and trabecular bone mineral content in women with endometriosis: effect of gonadotropin-releasing hormone agonist and danazol. Fert. Steril. 52,21,1989.
33. Shankar, G. and Stern, P. The hypothalamic hormones somatostatin and LHRH fail to affect bone resorption in vitro. ASBMR/ICCRH FIrst Joint Meeting, 1989.
34. Pasqualini, J.R., Sumida, C. and Giambiagi, N. Pharmacoldynamic and biologic effects of anti-estrogens in different models. J. Steroid Biochem. 31,613,1988.
35. Peplow, P. and Jeremy, J.Y. Antiestrogens: Structure, mode of action and relationship to eicosanoids. Prostag. Leukotr. and Essen. Fatty Acids 37,241,1989.
36. Richie, G.A.F. The direct inhibition of prostaglandin synthetase of human breast cancer tumor tissue by tamoxifen. Rec. Res. Canc. Res. 71,96,1980.
37. Turken, S., Siris, E., Seldin, D., Flaster, E., Hyman, G., Lindsay, R. Effects of tamoxifen on spinal bone density in women with breast cancer. J. Natl. Canc. Inst. 81,1086,1989.
38. Gotfredsen, A., Christiansen, C. and Palshof, T. The effect of tamoxifen on bone mineral content in premenopausal women with breast cancer. CAncer 53,853,1984.
39. Fentiman, I.S., Caleffi, M., Rodin, A., Merby, B. and Fogelman, I. Bone mineral content of women receiving tamoxifen for mastalgia. Brit. J. Cancer 60,262,1989.
40. Feldman, S., Minne, H.W., Parvizi, S., Pfeifer, M., Lempert, U.G., Bauss, F. and Ziegler, R. Antiestrogen and antiandrogen administration reduce bone mass in the rat. Bone and Mineral7:245,1989.
41. Beall, P.T., Misra, L.K., Young, R.L., Spjut, H.J., Evans, H.J. and LeBlanc, A. Clomiphene protects against osteoporosis in the mature ovariectomized rat. Calc. Tiss. Intl. 36,123,1984.
42. Turner, R., Wakley, G.K., Hannon, K.S. and Bell, N.S. Tamoxifen prevents the skeletal effects of ovarian hormone deficiency in rats. J. Bone Min. Res. 2,449,1987.
43. Wakley, G.K., Baum, B.L., Hannon, K.S. and Turner, R.T. The effects of tamoxifen on the osteopenia induced by sciatic neurotomy in the rat: a histomorphometic study. Calc. Tiss. Intl. 43,383,1988.

44. Goulding, A., Gold, E. and Fisher, L. Effects of clomiphene and tamoxifen in vivo on the bone-resorbing effects of parathyroid hormone and of high oral doses of calcitriol ($1,25(OH)_2D_3$) in rats with intact ovarian function consuming low calcium diet. Bone and Mineral 8,185,1990.
45. Stewart, P.J. and Stern, P.H. Effects of the antiestrogens, tamoxifen and clomiphene, on bone resorption in vitro. Endocrinology118,125,1986.
46. Rich, G.M., Mudge, G., and LeBoff, M.S. Cyclosporin A - associated osteoporosis in cardiac transplant patients. J. Bone Min. Res. 5, s183,1990.
47. Tugwell, P., Bombardier, C., Gent, M., Bennett, K.J., Bensen, W.G., Carette, S., Chalmers, A., Esdaile, J.M., Klinkhoff, A.V., Kraag, G.R., Ludwin, D., Roberts, R.S. Low-dose cyclosporin verses placebo in patients with rheumatoid arthritis. Lancet 335,1051,1990.
48. Schlosberg, M., Movsowitz, C., Epstein, S., Ismail, F., Fallon, M.D. and Thomas, S. The effect of cyclosporin A administration and its withdrawal on bone mineral metabolism in the rat. Endocrinology 124, 2179,1989.
49. Del Pozo, E. Graeber, M. Elford, P. and Payne, T. Regression of bone and cartilage loss in adjuvant arthritic rats after treatment with cyclosporin A. Arth. Rheum. 33,247,1990.
50. Skjodt, H. Crawford, A., Elford, P.R., Ihrie, E., Wood, D.D. and Russell, R.G.G. Cyclosporine A modulates interleukin-1 activity on bone in vitro. Brit. J. Rheumatol. 24, Suppl. 1, 165,1985.
51. Klaushofer, K., Hoffmann, O. Stewart, P.J., Czerwenka, E., Koller, K., Peterlik, M. and Stern, P.H. Cyclosporine A inhibits bone resorption in cultured neonatal mouse calvaria. J. Pharmacol. Exp. Therap. 243,584,1987.
52. Sasagawa, K., Fujibayashi, S., Okano, K., Nawa, C., Suzuki, S., Kuo, S., Yamada, Y. and Someya, K. Different inhibitory actions of immunomodulating agents and immunosuppressive agents on bone resorption of mouse calvaria. Int. J. Immunopharmacol. 11,953,1989.
53. Loertscher, R., Thiel, G., Harder, F. and Brunner, F.P. Persistent elevation of alkaline phosphatase in cyclosporine-treated renal transplant recipients. Transplantation 36,115,1983.
54. Hahn, T.J. Steroid and drug induced osteopenia, in Primer on the Metabolic Bone Diseases and Disorders of Mineral Metabolism, M.J. Favus, ed., American Society for Bone and Mineral Research, Kelseyville, 1990, 158.
55. Takeshita, N., Seino, Y., Ishida, H., Seino, Y., Tanaka, H., Tsutsumi, C., Ogata, K., Kiyohara, K., Kato, H., Nozawa, M., Akiyama, Y., Hara, K. and Imura, H. Increased circulating levels of γ-carboxyglutamic acid-containing protein and decreased bone mass in children on anticonvulsant therapy. Calc. Tiss. Intl. 44,80,1989.

PATHOGENETIC MECHANISMS OF SECONDARY OSTEOPOROSIS

Lawrence G. Raisz, M.D.
University of Connecticut Health Center
Farmington, CT 06030

I. INTRODUCTION

The term "secondary osteoporosis" is widely used to refer to those conditions in which decreased bone mass or strength is associated with a specific disease or pharmacologic agent. However, the incidence of osteoporotic fractures in the various forms of secondary osteoporosis is often low, except in that group which is most subject to what we term "primary osteoporosis", that is, in older individuals and particularly in postmenopausal women. Thus it is important to distinguish between a causal relationship between a disease or agent and osteoporosis and a less central role for that disease or agent as an aggravating factor. In considering pathogenesis, it is also important to recognize that the effects may be relatively indirect, due to an alteration in systemic hormone production or in the function of another organ. For example, the administration of gonadotrophin releasing hormone analogs which reduces gonadotrophin secretion and hence sex hormone production leads to bone loss by a relatively indirect mechanism. Phenytoin and other antiepileptic drugs may lead to bone loss by inhibiting calcium absorption in the intestine, which in turn may be related to a defect in the synthesis of the active forms of vitamin D. Finally, there are some recent data suggesting that secondary osteoporosis could be due to abnormalities in the production of local factors which regulate bone turnover. The number of local factors which have been identified as regulating bone metabolism is already large and is increasing. Thus pathogenetic mechanisms may be quite complex.

II. DISEASES AND FACTORS ASSOCIATED WITH SECONDARY OSTEOPOROSIS

Some of the diseases which are associated with secondary osteoporosis are listed in Table 1. Chronic liver disease, renal failure, malabsorption and epilepsy (with anti-epileptic drug therapy), have been associated with osteomalacia as well as osteoporosis. However, recent studies suggest that the most common abnormality of the skeleton in these disorders may be osteoporosis. The relative rarity of classical osteomalacia may be due to the fact that these patients are so often treated with a sufficient amount of vitamin D to enable them to mineralize their reduced skeletal mass. The reasons for the reduced skeletal mass are not well understood and are undoubtedly complex and multifactorial. For example, in alcoholism impairment in sex hormone metabolism, poor nutrition, decreased absorption of the minerals and direct effects on bone have all been implicated.

TABLE 1

Diseases Associated with Secondary Osteoporosis

1.	Cushing's syndrome	7.	Renal disease
2.	Hyperparathyroidism	8.	Malabsorption
3.	Hypogonadism	9.	Diabetes Mellitus
4.	Multiple myeloma	10.	Epilepsy
5.	Hyperthyroidism	11.	Mastocytosis
6.	Liver disease	12.	Hemolytic anemia

Some of the pathogenetic factors which may be important in secondary osteoporosis are listed in Table 2. They have been arbitrarily divided into those factors which are likely to produce osteoporosis when they are in excess and those which are likely to produce osteoporosis when they are deficient. Some of the factors listed are clearly important in secondary osteoporosis, such as glucocorticoids, parathyroid hormone and thyroid hormone, while for others the pathogenetic role is less clear. For example, it is well known that insulin is an important factor for the function of osteoblasts and a selective stimulator of collagen synthesis, but it is not clear that insulin deficiency is the cause of diminished bone mass or increased fractures in diabetic patients.[1] Indeed, the other consequences of insulin deficiency are so severe and the use of hormone replacement is so prevalent that deficiency may not occur for any length of time. Nevertheless, there is evidence that bone fractures are increased in diabetic patients.

TABLE 2

Potential Pathogenetic Factors in Secondary Osteoporosis

Excess	Deficiency
1. Glucocorticoids	1. Growth Factors
2. Parathyroid hormone	2. Insulin
3. Thyroid hormone	3. Sex hormones
4. Interleukins	4. Calcium
5. Prostaglandins	5. Phosphorus
6. Heparin	6. Vitamin C and D
7. Neuropeptides	7. Protein
8. Drugs	8. Trace elements

In this manuscript, I will discuss only a few of the diseases and agents listed in Tables 1 and 2, specifically glucocorticoids, parathyroid hormone, interleukins and heparin. Thyroid hormone, anticonvulsant drugs and other pharmacologic mechanisms for bone loss are discussed elsewhere in this volume by Dr. P.H. Stern.

III. GLUCOCORTICOIDS AND BONE

We have recently reviewed glucocorticoid-induced osteoporosis[2] and will only summarize the major points and add some recent information here. Even if glucocorticoids had no direct effect on the skeleton, one might expect to see osteoporosis based on indirect mechanisms (Table 3).

TABLE 3

Corticosteroids and Bone:
Indirect Mechanisms for Bone Loss

1. Decreased intestinal absorption of Ca and PO_4
2. Increased renal excretion of Ca and PO_4
3. Inhibition of gonadotropins
4. Loss of adrenal androgen (exogenous corticosteroids)
5. Muscle wasting
6. Immobilization due to weakness or associated disease

Glucocorticoids can decrease the absorption of calcium and phosphorus in the intestine and increase their renal excretion. At high concentrations, glucocorticoids can inhibit gonadotrophins and hence reduce sex hormone output. The most common cause of glucocorticoid-induced osteoporosis is exogenous administration of these hormones to patients with rheumatologic, pulmonary, gastrointestinal or dermatologic disorders. In these patients, the suppression of ACTH results in decreased adrenal androgen. In postmenopausal women, this adrenal androgen is the major source of sex hormones, largely by conversion to estrone in the peripheral adipose tissue. Finally, the diseases for which glucocorticoids are used often lead to immobilization and the glucocorticoids themselves can cause muscle weakness. The resultant decreased physical activity is likely to reduce bone mass.

Despite all these indirect mechanisms, glucocorticoid-induced osteoporosis is not universal and probably does depend more on direct mechanisms (Table 4). Secondary hyperparathyroidism has been described as an important component of glucocorticoid-induced osteoporosis but not all data support this. PTH levels are not always elevated and the pattern of bone loss differs from that associated with excess PTH (see below). Nevertheless, impaired calcium absorption is likely to lead to some increase in PTH production, and this in turn is likely to cause increased bone resorption. There is some recent evidence for direct stimulation of bone resorption by glucocorticoids.[3] The effect appears to be biphasic with stimulation only at low physiologic concentrations,

TABLE 4

Corticosteroids and Bone:
Direct Mechanisms for Bone Loss

1. Increased bone resorption
   - Secondary hyperparathyroidism
   - Direct stimulation of osteoclasts
2. Decreased bone formation
   - Inhibition of preosteoblast replication
   - Decreased IGF-1 production
   - Decreased PGE_2 production

and transient, so that bone resorption is inhibited with prolonged administration. The best explanation for this, which is analogous to the effects on bone formation described below, is that glucocorticoids at physiologic concentrations can produce a permissive enhancement of the function of osteoclast, but that at high concentrations they either inhibit the osteoclast themselves or prevent their formation by decreasing the replication and differentiation of osteoclast precursors.

The inhibitory effect of glucocorticoids on bone formation is probably the most important mechanism for the secondary osteoporosis so frequently encountered in patients treated with long-term steroids. Nevertheless, the effect on bone formation in vitro is biphasic. Glucocorticoids can increase the ability of osteoblasts to produce collagen, but with prolonged treatment collagen synthesis decreases.[2] This decrease is preceded by a decrease in DNA and protein synthesis in the periosteum, leading to the hypothesis that the anti-anabolic effect of glucocorticoids could be due to a decrease in the proliferation and/or differentiation of osteoblast precursors. This hypothesis has been amply confirmed in both cell and organ culture. It is supported by in vivo data on bone histomorphometry which showed that the mean wall thickness of packets of new bone is markedly decreased in glucocorticoid-treated patients, indicating that they are produced either by fewer or less active osteoblasts.[4]

The effect of glucocorticoids to inhibit DNA synthesis undoubtedly contributes to decreased bone formation, but is not essential for it. We found that when bones were treated with aphidicholin, a potent inhibitor of DNA polymerase, the inhibitory effect of glucocorticoids on collagen synthesis was still observed.[2] Subsequent studies showed that glucocorticoids decrease the production of IGF-1 by bone cells.[5] This decrease in IGF-1 production may well be a major pathogenetic mechanism since addition of IGF-1 can completely overcome the inhibitory effects of glucocorticoids.[6] We have also found that prostaglandin E_2 (PGE_2) can overcome the inhibition of both DNA and collagen synthesis by glucocorticoids.[7] However, here too the final common pathway may be IGF-1, since PGE_2 also increases IGF-1 production in bone.[8] In addition to their effects on IGF-1 and on cell replication, glucocorticoids can reduce the activity of transforming growth factor ß (TGFß) in bone cell cultures. This may also be a complex indirect pathway because glucocorticoids decrease the production of plasminogen activator which can in turn decrease the activation of latent TGFß.[9]

Thus it is likely that the anti-anabolic effects of glucocorticoids are complex and multifactorial. Whatever their mechanism, bone formation is clearly highly sensitive to these hormones. Indeed, small doses of prednisone which are no greater than those given for hormone replacement in Addison's disease may be sufficient to inhibit osteoblastic activity, as evidence by a recent study showing that glucocorticoids can block the nocturnal rise in osteocalcin in normal human subjects.[10] Thus the normal diurnal rhythm of glucocorticoid secretion with a decrease in the evening may be necessary to permit bone formation at night and sufficient to block bone formation during the morning.

IV. HYPERPARATHYROIDISM

The classic presentation of severe hyperparathyroidism with osteolytic lesions and multiple fractures is rarely encountered today. However, recent studies do indicate that there is an abnormality of bone in mild to moderate hyperparathyroidism.[11] This abnormality is quite different from the pattern of bone loss seen in primary osteoporosis or in glucocorticoid-induced osteoporosis. The major loss is in cortical bone with relative preservation of trabecular or cancellous bone. PTH has been shown to increase the production of IGF-1 and to activate TGFß in bone organ cultures[12,13] and it is possible that these growth factors are responsible for the maintenance of cancellous bone. The loss of cortical bone could then be ascribed to increased bone resorption, a direct effect of PTH. Why the cortical bone does not show the increase in formation that is seen in the cancellous bone is not clear. Possibly there is some defect in the ability to generate or respond to growth factors in the cortex.

Severe primary hyperparathyroidism can increase the incidence of both appendicular and axial fractures. Data on the incidence of fractures in milder cases are limited and there are studies suggesting that bone loss is not progressive in hyperparathyroidism.[14] Clinically, low bone mass is considered an indication for surgical treatment of primary hyperparathyroidism in the mild or asymptomatic case.

V. MULTIPLE MYELOMA

Since multiple myeloma most often causes localized lytic lesions, it is not generally considered as a cause of secondary osteoporosis. Nevertheless, patients who present with vertebral crush fractures are occasionally found to have multiple myeloma. In these individuals, there is presumably widespread distribution of the myeloma cells in the vertebral bodies. This presumably results in the local release of interleukin-1 (IL-1) and/or tumor necrosis factor ß (TNFß), products of myeloma cells, which are potent stimulators of bone resorption and inhibitors of bone formation.[15,16] The ability of cytokines produced in myeloma to mimic vertebral crush fracture syndrome lends some support to the hypothesis that cytokines are also pathogenetic factors in primary osteoporosis.[17] Recently, an IL-1 receptor antagonist has been identified which can completely block the effects of IL-1 on bone resorption.[18] Whether this agent will be useful in treating myeloma or other forms of osteoporosis remains to be seen.

VI. HEPARIN

Prolonged treatment with large doses of heparin has been shown to produce osteoporosis[19] and has been implicated in the association of osteoporosis and mast cell disease.[20] Mast cells have been found in the marrow of osteoporotic patients and osteoporosis is regularly encountered in patients with mastocytosis. However, mast cells produce many different products and whether or not heparin is the major culprit in the associated loss of bone remains to be determined. Clinically, heparin-induced osteoporosis is relatively uncommon, although the prolonged use of heparin during pregnancy in certain immune disorders may produce a new problem of bone loss. The mechanism by which heparin produces bone loss is still not fully understood. Direct stimulation of bone resorption has been reported, but we were unable to confirm this in a rat long bone organ culture system. We have observed that heparin will enhance the resorptive response to both acidic and basic FGF factors which are present in bone and could be responsible for increased resorption in the presence of excess heparin (Simmons, H.A. and Raisz, L.G., unpublished data). Finally, heparin has also been shown to inhibit bone formation.[21] This effect appears to be independent of its anti-coagulant activity and is seen with both the newer low molecular weight heparins as well as with the large molecular weight heterogeneous heparin preparations that have been used in the past.[22]

VIII. CONCLUSION

This brief summary of some of the pathogenetic mechanisms in osteoporosis certainly shows that the extent of our ignorance is greater than the extent of our knowledge. Most forms of secondary osteoporosis are not well understood and the specific roles of the many local factors in bone need to be elucidated before we can advance much further. Nevertheless, this field has been opened up by new discoveries concerning the local regulation of bone metabolism which should lead to important new information concerning pathogenesis and new approaches to therapy.

IX. REFERENCES

1. Melchior, T.M., Sorensen, O.H., Thamsborg, G., Sykulski, R., Storm, T. Bone tissue alterations in diabetes, in Tissue-Specific Metabolic Alterations in Diabetes: Frontiers in Diabetes, Vol. 10, Belfiore, F., Molinatti, G.M., Reaven, G.M., Karger, eds., 1990, pp. 40.

2. Lukert, B.P., Raisz, L.G. Glucocorticoid-induced osteoporosis: Pathogenesis and management. Ann. Int. Med. 112(5):352, 1990.

3. Gronowicz, G., McCarthy, M.B., Raisz, L.G. Glucocorticoids stimulate resorption in fetal rat parietal bones in vitro. J. Bone Min. Res. 5(12):1223, 1990.

4. Dempster, D.W. Perspectives: Bone histomorphometry in glucocorticoid-induced osteoporosis. J. Bone Min. Res. 4(2):137, 1989.

5. McCarthy, T.L., Centrella, M., Canalis, E. Cortisol inhibits the synthesis of insulin-like growth factor-I in skeletal cells. Endocrinol. 126(3):1569, 1990.

6. Kream, B.E., Petersen, D.N., Raisz, L.G. Cortisol enhances the anabolic effects of insulin-like growth factor I on collagen synthesis and procollagen messenger ribonucleic acid levels in cultured 21-day fetal rat calvariae. Endocrinol. 126:1576, 1990.

7. Raisz, L.G., Fall, P.M. Biphasic effects of prostaglandin E_2 on bone formation in cultured fetal rat calvariae: Interaction with cortisol. Endocrinol. 126(3):1654, 1989.

8. McCarthy, T.L., Centrella, M., Raisz, L.G., Canalis, E. Prostaglandin E_2 stimulates insulin-like growth factor I synthesis in osteoblast-enriched cultures from fetal rat bone. Endocrinol. 128(6), 1991.

9. Martin, T.J. Intercellular communications in bone, in Osteoporosis 1990, Christiansen, C. and Overgaard, K.,eds., 1990, pp. 221.

10. Nielson, F.H., Hunt, C.D., Mullen, L.M., Hunt, J.R. Effect of single doses of prednisone on the circadian rhythm of serum osteocalcin in normal subjects. J. Clin. Endocrinol. Metab. 67(5):1025, 1988.

11. Silverberg, S.J., Shane, E., de la Cruz, L. et al. Skeletal disease in primary hyperparathyroidism. J. Bone Min. Res. 4(3): 283, 1989.

12. Canalis, E., Centrella, M., Burch, W. et al. Insulin-like growth factor-1 mediates selective anabolic effects of parathyroid hormone in bone culture. J. Clin. Invest. 83(1):60, 1989.

13. Pfeilschifter, J., Mundy, G.R. Modulation of type ß transforming growth factor activity in bone cultures by osteotropic hormones. Proc. Nat. Acad. Sci. USA 84(7):2024, 1987.

14. Rao, D.S., Wilson, R.J., Kleerekoper, M. et al. Lack of biochemical progression of continuation of accelerated bone loss in mild asymptomatic primary hyperparathyroidism: Evidence for biphasic disease course. J. Clin. Endocrinol. Metab. 67(6):1294, 1988.

15. Kawano, M., Yamamoto, I., Iwato, K., Tanaka, H., Asaoku, H., Tanabe, O., Ishikawa, H., Nobuyoshi, M., Ohmoto, Y., Hirai, Y., Kuramoto, A. Interleukin-1-beta rather than lymphotoxin as the major bone resorbing activity in human multiple myeloma. Blood 73(6):1646, 1989.

16. Garrett, R.I., Durie, B.G.M., Nedwin, G.E., Gillespie, A., Bringman, T., Sabatini, M., Bertolini, D.R., Mundy, G.R. Production of lymphotoxin, a bone-resorbing cytokine, by cultured human myeloma cells. N. Engl. J. Med. 317(9):526, 1987.

17. Pacifici, R., Rifas, L., Teitelbaum, S., Slatopolsky, E., McCracken, R., Bergfeld, M., Lee, W., Avioli, L.V., Peck, W.A. Spontaneous release of interleukin 1 from human blood monocytes reflects bone formation in idiopathic osteoporosis. Proc. Nat. Acad. Sci. USA 84(13):4616, 1987.

18. Seckinger, P., Klein-Nulend, J., Alander, C., Thompson, R.C., Dayer, J.-M., Raisz, L.G. Natural and recombinant human IL-1 receptor antagonists block the effects of IL-1 on bone resorption and prostaglandin production. J. Immunol. 145(12):4181, 1990.

19. Ginsberg, J.S., Kowalchuk, G., Hirsh, J., Brill-Edwards, P., Burrows, R., Coates, G., Webber, C. Heparin effect on bone density. Thromb. Haemost. 64(2):286, 1990.

20. Chines, A., Pacifici, R., Avioli, L.V., Teitelbaum, S.L., Korenblat, P.E. Systemic mastocytosis presenting as osteoporosis - A clinical and histomorphometric study. J. Clin. Endocrinol. Metab. 72(1):140, 1991.

21. Hurley, M.M., Kream, B.E., Raisz, L.G. Effects of heparin on bone formation in cultured fetal rat calvaria. Calcif. Tiss. Int. 48:183, 1990.

22. Hurley, M.M., Kream, B.E., Raisz, L.G. Structural determinants of the capacity of heparin to inhibit collagen synthesis in 21-day fetal rat calvariae. J. Bone Min. Res. 5(11):1127, 1990.

RISKS OF DRUG TREATMENTS OF METABOLIC BONE DISEASES

R. Ziegler, Dept. of Internal Medicine I (Endocrinology and Metabolism), University of Heidelberg, Heidelberg/Germany

INTRODUCTION

Already Theophrastus Bombastus of Hohenheim (1493-1541), called Paracelsus, recognized that effects and side effects may be hidden in the same agent: "Everything is a poison and nothing is without poison. It is only the dosage that a thing is no poison".

This truth is also valid for drugs which are used for metabolic bone diseases, and the present review deals with important representatives of such drugs used for the treatment of endocrine and metabolic osteopathies. The guideline is the pathophysiological scheme of figure 1 where drugs for the treatment of osteoporosis are enlisted in the respective positions of metabolic steps. Of course, the use of single drugs for other diseases will be also mentioned, but osteoporosis is in fact the disease in which at the moment the largest number of agents is administered.

For direct substitution of the extinguished ovarian function, of course **estrogens** are used. **Anabolics** may serve in a comparable substitutive way, but it is not excluded that they may exert additional effects e.g. on muscle.
According to some authors, calcitonin secretion is diminished in osteoporotic women (1). Therefore, **calcitonin** is tried for the treatment of osteoporosis, be it as a substitutional or a pharmacological treatment. Pharmacological doses of calcitonin are most efficient in Paget's disease of bone.
There is (eventually) not only a fall in calcitonin, but also **parathyroid hormone (PTH)** is lowered after menopause. Presumably this is due to the small increase in serum calcium which is especially evident in an acute manner after oophorectomy (2). Trials using PTH for the treatment of osteoporosis are going on.
PTH is not only a stimulator for bone turnover, but it also acts as a glandotrophic hormone in the endocrine gland "kidney" where it stimulates the formation of **1,25-dihydroxycholecalciferol = calcitriol.** The lack of calcitriol leads to diminished calcium absorption, and at the same time more **calcium** is lost through the kidney because serum calcium is increased and PTH (normally decreasing calciuria) is lowered. Therefore, the so-called

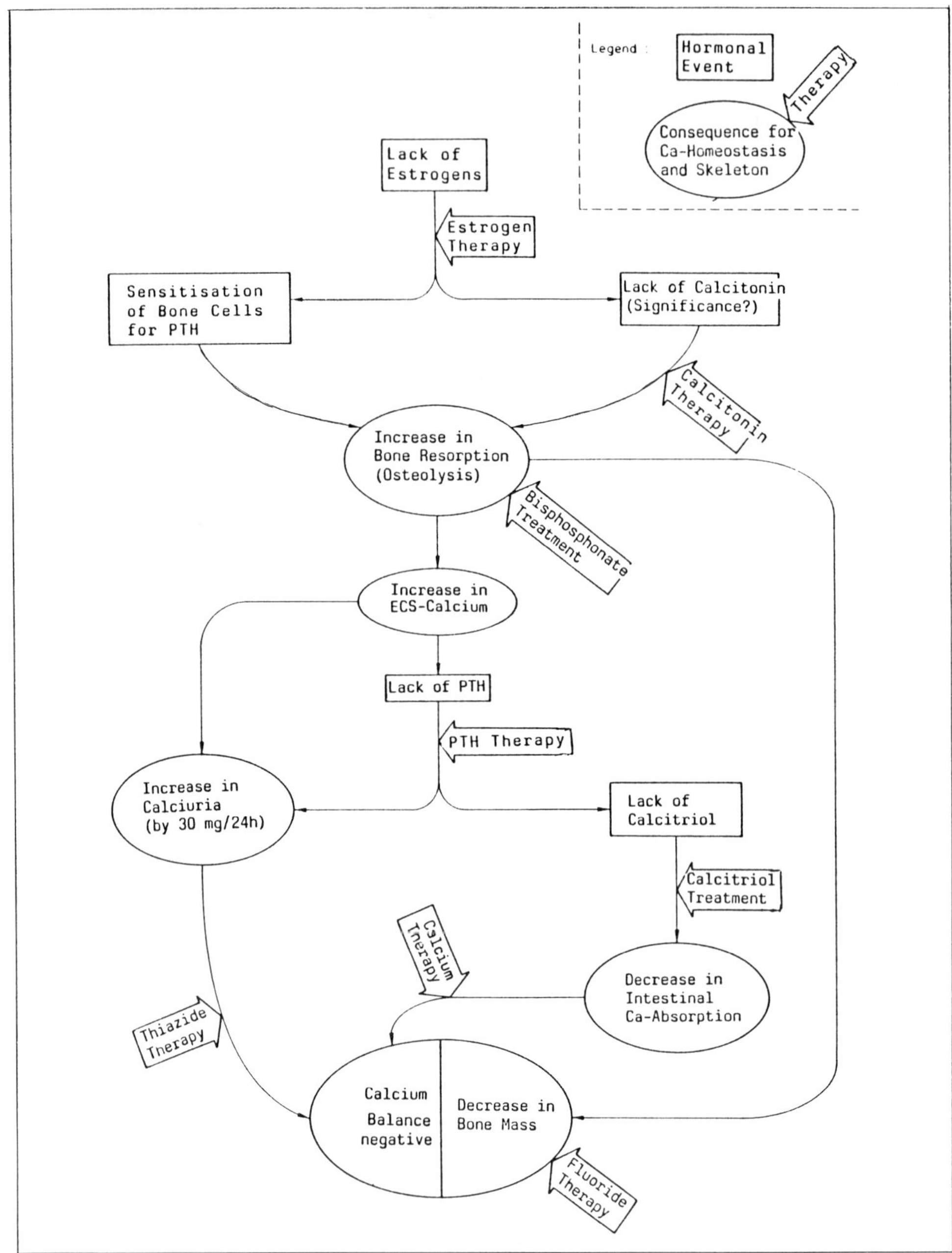

FIGURE 1
Drugs, used for the treatment of osteoporosis, and the steps of pathogenesis of osteoporosis where they interact

"calcium treatment" with or without vitamin D and metabolites is a kind of basic or standard substitutional treatment of osteoporosis.

Osteoporosis is characterized by a dysbalance between bone resorption and bone formation, the latter not being adequate to the former. Therefore, antiresorptive principles are considered for treatment. As a physiological factor inhibiting osteoclasts, calcitonin already has been mentioned. But still higher antiosteoclastic activity is exerted by the **bisphosphonates.** These pharmacological substances are analogues of pyrophosphate which is too labile to be used for therapeutic purposes. Bisphosphonates are most important drugs for hypercalcemia of malignancy and Paget's disease of bone.

Finally, **fluorides** are used as stimulators of osteoblasts in order to increase bone mass in osteoporosis. The discussion on the efficiency of this kind of treatment has not yet come to an end (3).

SEX HORMONES

In both sexes, hypogonadism leads to diminished bone mass - substitutional treatment restores the defect. The difference between the sexes is the menopause: in the natural course, men keep their gonadal hormones after the fifties, but women don't.

MEN:
If a man develops hypogonadism, he should be adequately substituted with testosterone. This infers the "natural" risk of prostate cancer which does not develop in hypogonadal men (4). Thus, testosterone substitution e.g. for osteoporosis prophylaxis may be accompanied by restoration of this "natural" risk.

There are no data whether e.g. anabolics maintain bone mass without influencing the prostate gland; as they suppress (at least in the dosages of misuse for body building) the gonadotrophins, in males with normal testes these organs shrink and sexual function is impaired.

WOMEN:
Besides many minor changes in well-being and behaviour connected with estrogen replacement therapy (ERT) (5), the respective women are aware of three major problems which may show up: 1) Changes in body weight, 2) cardiovascular disease, 3) malignancies (Table 1).

TABLE 1
Side effects of estrogen (plus progestagen) substitution after the menopause

Estrogen (+ Gestagen) Substitution in Postmenopausal Women	
Side Effects: + = positive, - = negative	
Fluid retention	+ less dryness of mucosae - weight gain
Lipoproteins (Cardiovascular risk)	+ estrogens ? estrogens plus progestogens
Endometrial cancer	- = increased for estrogens **alone** + = decreased risk for estrogen+progestogen
Mammary Cancer ?	

Ad 1: The hypoestrogenic state leads to a certain fluid loss which contributes to the genital dryness, small joint pain and others. When estrogen substitution is initiated, this may lead to fluid retention of 1 or even 2 liters. Some women in addition continuously gain weight, presumably by changing eating habits. Warnings and dietary recommendations are useful.

Ad 2 + 3: There is no doubt that "pure" estrogen substitution leads to a reduction in arteriosclerotic cardiovascular disease (CVD), like shown by Bush et al. (6): CVD mortality fell from 30.2 (in nonusers) to 12.8 per 10 000 patient years in ERT-users. In case of existing CVD at baseline, the fall was even more dramatic from 66.3 to 13.8. However, most of such studies documenting a protective estrogen effect stem from times of estrogen-monotherapy which was switched to the combination with progestogens in order to diminish the obvious risk of endometrial cancer which in fact was achieved. E.g. Gambrell (7) reported the incidence of 2.0 endometrial carcinomas in untreated women; women taking estrogens alone showed an incidence of 3.8 which fell to 0.3 for women substituted with estrogen plus progestogen. Similar data were published by Hammond et al. (8). Therefore, the addition of progestogens is mandatory for ERT in women who are not hysterectomized, but they may lose the profit for the cardiovascular system: "unfavorable" progestogens may neutralize the positive estrogen effect on blood lipids by lowering HDL cholesterol (9). This negative effect of some progestogens leading to increased arterial disease in women e.g. receiving norethindrone acetate alone (10) may be prevented by the selection of "favorable" progestogens for the combination with estrogens. The respective epidemiological studies of the outcome are urgently needed.

With respect to the incidence of mammary cancer in women taking ERT, a protective effect (relative risk of 0.68) was reported for the combination of estrogens plus progestogens as well as for estrogens alone (Table 2).

TABLE 2
Relative risk for breast cancer according to several reports

Mammary Cancer and ERT		
Gambrell et al. (1983)[11]	RR o.68	
Hunt et al. (1987) [12]	RR 1.59	for breast cancer
	RR o.55	for breast cancer mortality
Bergkvist et al. (1989)[13]	RR 1.1	(all cases = 2o8)
Subgroups:	RR 1.7	(E longer than 9 years) (23 cases)
	RR 4.4	(E plus progestogens longer than 6 years) 1o cases)

Hunt et al. (12) found a mammary cancer relative risk of 1.59 (related to the cancer statistics), but mammary cancer mortality had a RR of only 0.55. Most of the affected were grade I due to early recognition and control examinations during ERT.

In 1989, Bergkvist published new data indicating a rather high relative risk of 4.4 for women taking estrogens plus progestogens longer than 6 years. However, the statistics are to be debated and the respective subgroup is rather small (n = 10). Therefore, the reevaluation of this study is mandatory.

When discussing ERT it has to be kept in mind that compliance is far from being ideal and that the acceptance by the doctors themselves needs improvement: too many women stop ERT because of advices from their doctors (54.5 %) or because of complications (25 %) (see Table 3, according to Limouzin-Lamothe, 14). The differences between countries are remarkable.

In summary, side effects of ERT need further exploration to be sure that there is no cancer risk and that the optimal progestogen is added. But nevertheless, at the moment ERT can be recommended as a useful treatment with **minor** side effects.

TABLE 3
Reasons for stopping hormonal replacement therapy, according to Limouzin-Lamothe (14)

	USA	France	Italy	Cumulative
Doctor's advice	54,5	46,2	66,7	54,5
Felt better, no longer needed	34,1	34,6	11,3	29,5
Felt worse	6,8	3,8	-	4,5
Didn't feel better	9,1	11,5	16,7	11,1
Complications	31,8	23,1	11,1	25,0
Too expensive	0	0	0	0
duration of treatment	6,3	5,2	3,2 years	
Awareness of osteoporosis	95,7%	57,7%	87%	

If **anabolics** are administered to postmenopausal women, negative side effects are to be expected in many cases (Table 4).

Voice changes are seen up to more than half of the treated in case of higher doses (15), but prolongation of the injection intervals diminishes the problem. Increased hair growth (which is often an already inherent problem for postmenopausal women) happens in 20 to 30 %. Fluid retention and weight gain, sometimes leading to ankle edema, are further events. The authors of the cited studies (15, 16, 17, 18) state that the side effects were often reversible and only rarely motivated the women to stop treatment. More severe complications known from very high doses were not seen. Whether the potential positive side effect of muscular hypertrophy contributes to the therapeutic goal of improving bone mass, is not yet known.

Calcitonin

With respect to calcitonin treatment, three questions are to be answered: (a) Are there organ lesions? (b) Are there relevant functional disturbances? (c) Are there adverse reactions?

TABLE 4
Side effects of treatment of osteoporosis in women using anabolics (according to 15, 16, 17, 18)

Anabolics				
Negative Side Effect	Chesnut et al. 1983 [16] n = 23	Gennari et al. 1989 [17] n = 10	Hassager et al. 1989 [15] n = 25	Need et al. 1989 [18] n=70
Voice changes		2/10	11/14 (3 weekly) 4/11 (4 weekly) 5/25	
Increased hair growth	30 %			
Increase AST	43 %			
Ankle edema	22 %			
Weight gain				+1,5 kg (n.s.)
Acne	9 %			
Requiring higher doses: Peliosis hepatis, malignant hepatoma, relevant hyperlipoproteinemia				
Potential **positive** side effect: Increase in muscle mass				

Ad (a): It is one of the important advantages of calcitonin treatment that there are no lesions in any organ of the body due to longterm treatment. This is due to the fact that calcitonin is not only a natural hormone, but the organism also escapes from reacting on chronically elevated calcitonin levels. When patients with medullary thyroid carcinoma presenting with extremely high endogeneous calcitonin levels were investigated, no lesions in important organs like liver, kidney, bone marrow a.s.f. were discovered; the same is true for patients with Paget's disease of bone during long term treatment (19).

Ad (b): This question was extensively studied e.g. in gastrointestinal side effects (20, 21). When acutely administered, calcitonin inhibits insulin secretion and impairs glucose tolerance; this could have inferred the risk of diabetogenicity. Furthermore, calcitonin inhibits gastric secretion as well as pancreatic enzyme secretion. However, the examination of the functional capacity of the stomach as well as the exocrine and endocrine pancreas revealed that calcitonin lost its inhibitory potency during chronic hypercalcitoninism, be it from C-cell carcinoma or from therapeutic administration. After calcitonin withdrawal, the escape phenomenon vanishes, and the reactivity of the respective organs is restored. Presumably, also the bone cells may be subjected to the escape phenomenon (with the exception of the diseased osteoclasts of Pagetic bone areas), and this should be kept in mind in case a bone disease without a marker of calcitonin activity is treated.

Ad (c): It is well known that calcitonin may induce some adverse reactions (Table 5).

TABLE 5
Adverse reactions after calcitonin administration
(according to 22 and 23)

Gastrointestinal symptoms:
- Nausea
- Vomiting
- Abdominal pain, cramps
- Diarrhoea
- Unpleasant metallic taste in the mouth

Vascular symptoms:
- Facial flushing, reddening of ears
- Sensation of facial warmth
- Sensation of warmth affecting the hands
- Tingling in the extremities

Local symptoms:
- Erythema at the site of injection
- Pain at the site of injection

Renal symptoms:
- Increased frequency of micturition
- Polyuria

Allergic symptoms:
- Rash

Immune reactions:
- Neutralizing antibodies

Patients may be irritated by flushs (reddening of the face, ears, and sometimes the breast). They may feel nauseous and also vomiting rarely occurs. Abdominal complaints may be cramps, diarrhea or polyuria.

The severity of such side effects is on one side in part dose dependent, but has on the other side to be differentiated from placebo effects which accompany any kind of treatment. This can be seen in a study from Ringe (24) who found e.g. the report of nausea in 8 out of 20 control patients receiving calcium alone whereas the additional administration of calcitonin only produced nausea in 5 out of 19 patients receiving calcitonin every 2nd day and in 7 out of 20 patients on daily calcitonin. However, flushs and hot feelings were only reported by calcitonin treated patients (lower dose: 4/19, higher dose 5/20).

Presumably, the diagnosis may be also relevant for the occurrence and registration of such side effects: Caniggia (25) using carbocalcitonin observed flushing in 12 % of patients with Paget's disease, in 3,5 % of patients with Sudeck's syndrome, but in no case of osteoporosis.

There are some hints that the pattern of subjective adverse reactions could be different depending on the calcitonin species which is used. Eisinger and Ouaniche (26) found a trend to more gastrointestinal side effects after human calcitonin (HCT) than after salmon calcitonin (SCT), but vasomotor side effects were equal in 25 patients. In 10 healthy volunteers HCT produced more and longer lasting flushes. In an own retrospective analysis of 39 patients with Paget's disease of bone who were treated by different regimens (mean: 2.4 treatment schedules per patient), the reports of side effects were not different for HCT and SCT (Figure 2). Possibly SCT given as a nasal spray produces less side effects, but before definite conclusions can be drawn exact absorption rates have to be measured: in case of HCT by nose we only saw an absorption of about 10 %; when giving SCT by nose in Paget's disease of bone, 400 IU nasally were not yet as efficient as 100 IU subcutaneously (27).

When comparing the modified eel-calcitonin carbocalcitonin with SCT and HCT in patients with Paget's disease, obviously less side effects are produced by carbocalcitonin (e.g. nausea-vomiting in 2.8 % versus 11.4 % or flushing in 11.4 % versus 23 %) (25).

Part of the side effects refers to local complaints at the injection site - here, the calcitonin species seems to be less important than the composition of the vehicle: when comparing a SCT (in acetic acid) and a HCT preparation (dissolved in mannit), HCT was better tolerated (28).

Finally, antibody formation against non-human calcitonins has to be mentioned. Antibodies without biological consequences are rather frequent (40 to 70 %; see 29), but clinical resistance during e.g. SCT treatment happens less often (5 to 40 %; 30, 31). If a disease has a biochemical marker of calcitonin efficiency (like alkaline phosphatase in Paget's disease of bone), neutralizing antibodies will be easily recognized; but in diseases without such a marker undiscovered formation of neutralizing antibodies may be misdiagnosed as non-response.

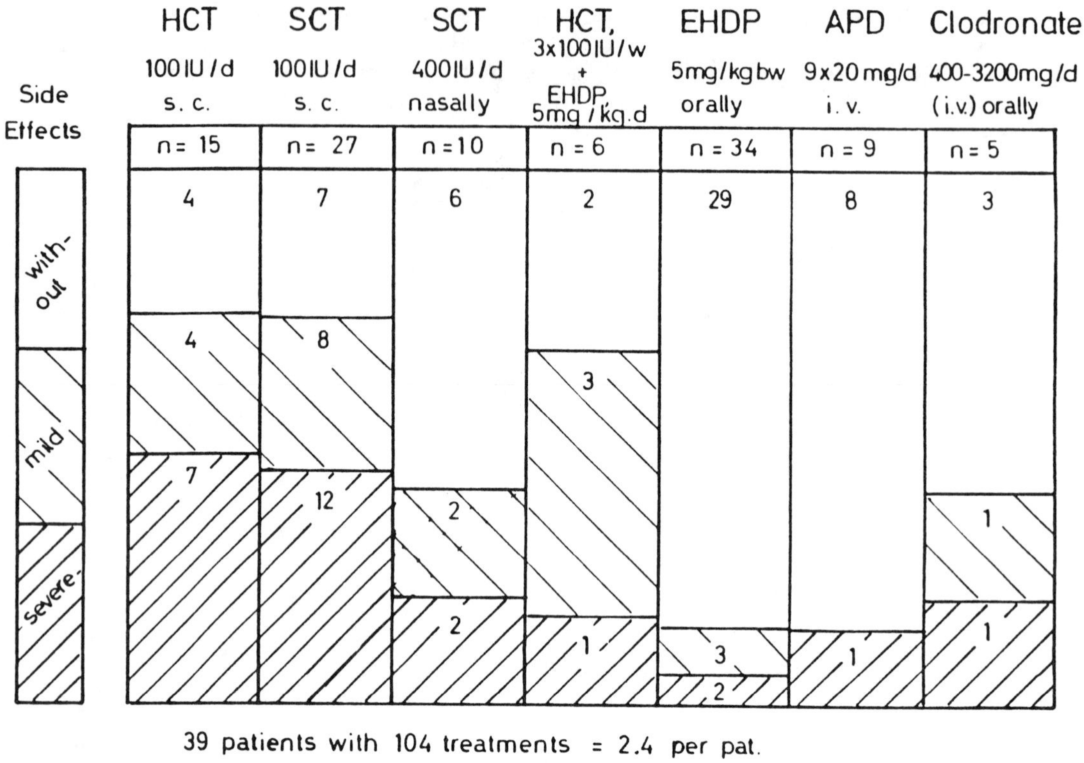

Side Effects	HCT 100IU/d s.c.	SCT 100IU/d s.c.	SCT 400IU/d nasally	HCT, 3x100IU/w + EHDP 5mg/kg.d	EHDP 5mg/kgbw orally	APD 9x20 mg/d i.v.	Clodronate 400-3200mg/d (i.v.) orally
	n= 15	n= 27	n =10	n = 6	n = 34	n = 9	n = 5
with-out	4	7	6	2	29	8	3
mild	4	8	2	3	3		1
severe	7	12	2	1	2	1	1

39 patients with 104 treatments = 2.4 per pat.

FIGURE 2
Report of side effects in a retrospective interrogation of 39 patients with Paget's disease of bone receiving different treatment schedules (mean: 2.4) over the years. The columns represent the respective subcollectives (numbers of patients on top), the encircled numbers how the patients' distribution according to no side effects, mild, or severe side effects.

Parathyroid hormone (PTH)
PTH is only used in a few studies - not all mention the looking for side effects. Presumably, relevant hazards were not observed. In an open study in 8 women using PTH 1-38 as an activitor (according to the ADFR-concept, 32) and EHDP as a depressor, in two patients a blood pressure increase accompanied by flush-like symptoms was observed. There was no evidence of PTH-antibody formation (33).

Calcitriol and Calcium
Lowered PTH leads to reduced calcitriol formation (Figure 1), due to different degrees of hypovitaminosis D calcium absorption is lowered in osteoporosis. Therefore, calcium as

well as calcitriol or combinations of both are used for treatment.

Data on side effects of pure calcium treatment are rare, but the effects of a pure calcitriol or a combined treatment are better documented. In 1985, Riggs and Nelson (34) treated 48 osteoporotics with 0.5 to 0.75 μg calcitriol per day. They did not see renal lithiasis nor significant clinical complications. 7 cases had mild hypercalcemia, and 8 presented with hypercalciuria (>350 mg/day). The dose had to be reduced in one quarter of the treated. Corresponding observations were made by Ott and Chesnut (35) who applied 1000 mg calcium and 0.43 μg calcitriol (mean) per day to 43 osteoporotics. There were no major clinical side effects, but peak calciuria up to 15 mmol/day as well as episodic hypercalcemia in 8 patients (Table 6). Therefore, especially the combination of calcitriol and calcium requires exact monitoring of calcemia and calciuria.

TABLE 6
Treatment of osteoporosis with calcitriol and calcium: side effects (according to 34, 35)

Calcitriol (o.5o - o.75 μg/d)

Riggs & Nelson 1985 [34] 48 pat.

... no renal lithiasis or significant clinical complications ...

... 7 patients with mild hypercalcemia ...

... 8 patients with hypercalciuria (>35o mg/day) ...

... reduction of dose in one quarter of patients ...

... 6 patients discontinued because of intercurrent illness ...

Calcium (1ooo mg/d) plus **Calcitriol** (mean o.43 μg/d)

Ott & Chesnut, 1989 [35] 43 pat.

... no major clinical side effects ...

... peak calciuria: 15 mmol/d ...

... sustained calciuria: < 8.7 mmol/d ...

... episodic hypercalcemia in 8 pat. ...

Among the agents discussed for the treatment of osteoporosis, also for calcium and calcitriol potential positive side effects have to be mentioned: diminished calcium intake could be related to arterial hypertension (36), and optimized calcium supply perhaps could not only protect bone but also the cardiovascular system. Furthermore, calcium supply in the food is discussed for the prophylaxis of colonic cancer in affected families (37). Supporting a connection between colon cancer and calcium homeostasis, Garland et al. (38) found lower 25-OH-vitamin D concentrations in respective cancer patients. Optimizing calcium and vitamin supply could, therefore, be of importance not only for the maintainance of the skeleton but also protect against colon cancer.

Bisphosphonates

In contrast to the until now discussed agents which are physiological factors and substrates of calcium homeostasis, the bisphosphonates are pharmacological substances which are used because of their antiosteoclastic activity. They are used with excellent efficiency in hyperosteoclastic diseases like Paget's disease of bone as well as humoral hypercalcemia of malignancy (HHM), and they also look promising for the treatment of osteoporosis (39, 40, 41, 42).

Bisphosphonates are poorly absorbed in the gut (only 2-5 %) (43); therefore larger quantities of the drug may cause gastrointestinal complaints. In patients with preexisting symptoms (e.g. nausea due to hypercalcemia), the intravenous bisphosphonate administration is a valuable alternative. Slow speed infusion is required because as a bolus, bisphosphonates may be deleterious for kidney function.

Etidronate/EHDP: This "first generation" bisphosphonate is very well tolerated and without relevant side effects when given in a dose of 5 mg/kg body weight per day (up to 15) (44). An increased number of fractures only is to be expected if higher doses (10 or 20 mg/kg body weight and day) are used over prolonged periods (9 to 24 months) (45).

When using much smaller doses cyclically in osteoporosis, no significant side effects were seen (41). In a study using also phosphate, diarrhea was seen during the phosphate phases in 39 % whereas placebo as well as EHDP phases were accompanied by diarrhea in 7 to 9 % of cases (42).

With respect to liver function, kidney function or bone marrow, no side effects due to EHDP-treatment were seen.

Pamidronate, APD: This "second generation" bisphosphonate is a potent osteoclast inhibitor without impairing mineralization. After initiating APD treatment, some patients may develop increases in body temperature after a few days for

48 hours or less; leucocytes may be lowered during this time. Definite leucopenia has not been observed. Gastrointestinal tolerance is dose dependent: When 600 mg/daily were given to cancer patients, a quarter of the patients dropped out because of nausea and vomiting (46). This rate of drop-out fell to 8 % after lowering the dosage to 300 mg/day. No other toxicity was seen (47).

Clodronate: This very potent "second generation" bisphosphonate was banned for use for about one decade after single cases of leukemia had been observed among clodronate treated patients; however, further observations and exact case analyses did not confirm a causal connection and no additional cases showed up (48).

When clodronate was given to patients with HHM, tolerance was very good: in 3 out of 34 cases gastrointestinal complaints were seen leading to discontinuation of the treatment in one case. The function of vital organs was not affected (49).

Fluoride

Fluoride treatment of osteoporosis is performed since more than 20 years, but data of controlled and even randomized studies mostly stem from recent years.

Franke and Runge (50) presented a retrospective analysis of 158 patients treated with three different NaF preparations. Gastrointestinal complaints were observed in 44 % of patients receiving pure NaF whereas gastric juice resistant preparations induced less complaints (32.9 % in case of moderate resistance, 27.0 % in case of strong resistance) (Table 6).

The mean value of 34.2 % is in accordance with the data of Mamelle et al. (51): they found more than one episode of gastrointestinal disorders in 39.4 % of their 180 NaF-treated patients. However, the frequency of the same complaints was 44.1 % in 136 patients with non-NaF-treatment indicating that the side effects of alternative schedules have to be considered, too. Riggs et al. (52) saw a relative risk of gastric pain, nausea, or vomiting in their NaF + calcium-group of 2.88 (compared with treatment with calcium alone). In Pak's patients (53), gastrointestinal side effects were less frequent.

The most intensively discussed side effect of fluorides are arthralgias, also called the "lower extremity pain syndrome". Whether the term "microfractures" or "incomplete fractures" (52) is justified, is unclear at the moment; however, as those regions never lead to mechanical damage, the term fracture is perhaps misleading.

TABLE 6
Treatment of osteoporosis with fluorides: side effects (according to 50, 51, 52, 53)

Fluoride treatment / Side effects

Franke & Runge 1987 [50] 158 pat.

Preparation	Gastrointestinal Complaints %	Arthralgia %	Gastric Ulcer %	No Side Effects %
Pure NaF	44.0	47.4	2.6	34.2
Moderate Gastric Juice Resistance	32.9	48.4	2.4	26.8
Strong Gastric Juice Resistance	27.0	67.6	10.8	29.7
All	34.2	52.5	4.4	29.1

Mamelle et al. 1988 [51] 180 pat. Non-NAF: 136 pat.)

Treatment	≥1 episode of gastrointestinal disorders %	≥1 episode of osteo-articular pain %	≥1 episode of other disorders (flushes, pruritus etc.) %
NaF	39.4	37.2	7.8
Non-NaF	44.1	30.1	13.2

Riggs et al. 1990 [52]

Fluoride: 310 patient-years Placebo: 325 patient-years

Relative Risk (RR)	Gastric pain, Nausea, Vomiting	Gastrointest. bleeding, anemia	Lower extremity pain	Bone spurs
Fluoride + Ca / Placebo + Ca	2.88	0.87	9.85	0.79

Pak et al. 1989 [53] 44 pat.

Nausea %	Abdominal pain %	Occult blood %	Foot pain %	Hip pain %	Generalized joint pain %
~10	2.4	11.4	9.2	2.3	2.3

The pain syndrom is seen in a frequency of about 10 % (53) up to 50 % (50); in Riggs' study (52) the relative risk (RR) is 9.85. Again Mamelle's data (51) are worth to be mentioned showing more than one episode of osteo-articular pain in 37.2 % of cases of fluoride treatment, but also in 30.1 % of alternative treatment schedules.

Whether complete fractions happen more often during fluoride treatment than expected (54), needs to be explored further.

Some of the fluoride side effects may be in part prevented by chosing lower doses (connected with the risk of diminished efficiency?) or other galenic preparations like monofluorophosphates which perhaps are better tolerated by the gastrointestinal tract (55).

CONCLUSIONS

There is no potent osteotropic or calciotropic agent which is free of side effects. For every indication, the spectrum of side effects even of the same agent, may be somewhat different. Differential treatment especially of osteoporosis should include considerations of potential side effects of the drug of choice as well as of the alternatives. The personality of the patient may be an additional factor determining compliance as well as the intensity of side effects.

REFERENCES

1) MacIntyre, I., Stevenson, J.C., A modern view of its physiological role and interrelation with other hormones, in Calcitonin 1980, Pecile, A., Ed., Excerpta Medica, Amsterdam Oxford Princeton, ICS, 540, 1981. p. 1.

2) Gallagher, J.C., Nordin, B.E.C., Oestrogens and calcium metabolism, in Oestrogens and ageing, van Keep, P.A. and Lauritzen, C., Eds., Karger, Basel, 1973.

3) Ziegler, R., Quo vadis der Osteoporose-Behandlung - mit neuen Daten aus der Klemme?, Therapiewoche, 40, 693, 1990.

4) Huggins, C., Hodges, C.V., Studies on prostatic cancer. I. The effect of castration, of estrogen and of androgen injection on serum phosphatates in metastatic carcinoma of the prostate, Cancer Res., 1, 293, 1941.

5) Sarrel, P.M., Estrogen replacement therapy, Obstet. Gynecol., 72, 2S, 1988.

6) Bush, T.L., Barrett-Connor, E., Cowan, L.D., et al., Cardiovascular mortality and noncontraceptive use of estrogen in women: Results form the Lipid Research Clinics Program Follow-Up Study, Circulation, 75, 1102, 1987.

7) Gambrell, R.D.,Jr., The prevention of endometrial cancer in postmenopausal women with progestogens, Maturitas, 1, 107, 1978.

8) Hammond, C.B., Jelovsek, F.R., Lee, K.L., Creasman, W.T., Parker, R.T., Effects of long-term estrogen replacement therapy. II. Neoplasia, Am. J. Obstet. Gynecol., 133, 537, 1979

9) Fähraeus, L., Larsson-Cohn, U., Wallentin, L., L-norgestrel and progesterone have different influences on plasma lipoproteins, Eur. J. Clin. Invest., 13, 447, 1983.

10) Kay, C.R., Progestogens and arterial disease - evidence from the Royal College of General Practitioners' study, Am. J. Obstet. Gynecol., 142, 762, 1982.

11) Gambrell, R.D., Maier, R.C., Sanders, B.I.: Decreased incidence of breast cancer in postmenopausal estrogen-progestogen users, Obstet. Gynecol., 62, 435, 1983.

12) Hunt, K., Vessey, M., McPherson, K., Coleman, M., Long-term surveillance of mortality and cancer incidence in women receiving hormone replacement therapy, Br. J. Obstet. Gynecol., 94, 620, 1987

13) Bergkvist, L., Adami, H.-O., Persson, I., Hoover, R., Schairer, C., The risk of breast cancer after estrogen and estrogen-progestin replacement, New Engl. J. Med., 321, 293, 1989.

14) Limouzin-Lamothe, M.-A., A review of patients' attitudes towards osteoporosis and its treatment (Italy, France and USA), in New Horizons in Osteoporosis, Christiansen, C., Ed., The Parthenon Publishing Group, Carnforth, 1988, p. 43.

15) Hassager, C., Riis, B.J., Pødenphant, J., Christiansen, C., Nandrolone decanoate treatment of post-menopausal osteoporosis for 2 years and effects of withdrawal, Maturitas, 11, 305, 1989.

16) Chesnut, C.H., Ivey, J.L., Gruber, H.E., Matthews, M., Nelp, W.B., Sisom, K., Baylink, D.J., Stanozolol in postmenopausal osteoporosis: therapeutic efficacy and possible mechanisms of action. Metabolism, 32, 571, 1983.

17) Gennari, C., AgnusDei, D., Gonnelli, S., Nardi, P., Effects of nandrolone decanoate therapy on bone mass and calcium metabolism in women with established post-menopausal osteoporosis: a double-blind placebo-controlled study, Maturitas, 11, 187, 1989.

18) Need, A.G., Horowitz, M., Walker, C.J., Chatterton, B.E., Chapman, I.C., Nordin, B.E.C., Cross-over study of fat-corrected forearm mineral content during nandrolone decanoate therapy for osteoporosis, Bone, 10, 3, 1989.

19) MacIntyre, I., Ed., Human calcitonin and Paget's disease, Hans Huber, Bern Stuttgart Vienna, 1977.

20) Ziegler, R., Raue, F., Holz, G., Minne, H., Calcitonin and the endocrine pancreas, in Calcitonin 1980, Pecile, A., Ed., Exerpta Medica, Amsterdam, ICS, 540, 1981, p. 314.

21) Hotz, J., Goebell, H., Ziegler, R., Magensekretion und exokrine Pankreasfunktion bei Patienten mit medullärem Schilddrüsenkarzinom sowie mit Morbus Paget unter Therapie mit Calcitonin, Z. Gastroenterol., 16, 547, 1978.

22) Azria, M., The Calcitonins. Physiology and Pharmacology, Karger, Basel, 1989.

23) Eisinger, J., Ouaniche, J., Side effects of different calcitonins, in Abstracts for "Calcitonin in osteoporosis", Int. Symp. Naxos, Sandoz SpA, Milan, 1986, 54.

24) Ringe, J.D., Behandlung der primären Osteoporose mit Calcium und Lachscalcitonin. DMW, 115, 1176, 1990.

25) Caniggia, A., Ed., Carbocalcitonin. The first ethylene-bridge calcitonin, Clin. Trials J., 23, 119, 1986.

26) Ziegler, R., Holz, G., Raue, F., Streibl, W., Nasal application of human calcitonin in Paget's disease of bone, in Molecular Endocrinology, MacIntrye, I., Szelke, M., Eds., Elsevier, 1979. 293.

27) Ziegler, R., Morbus Paget des Skelettes, Internist, 31, 763, 1990

28) Wüster, Chr., Schurr, W., Scharla, S.H., Raue, F., Minne, H.W., Ziegler, R., Improved local acceptance of human versus salmon calcitonin preparations in young healthy volunteers, submitted for publication, 1990.

29) Grauer, A., Raue, F., Schneider, H.G., Frank-Raue, K., Ziegler, R., In vitro detection of neutralizing antibodies after treatment of Paget's disease of bone with nasal salmon calcitonin, J. Bone Min. Res. 5, 387, 1990.

30) Woodhouse, N.J.Y., Mohamedally, S.M., Saed-Nejad, F., Martin, T.J., Development and significance of antibodies to salmon calcitonin in patients with Paget's disease on long-term treatment, Br. Med. J., 2, 927, 1977.

31) Rojannasathit, S., Rosenberg, E., Haddad, J.G., Jr., Paget's bone disease. Response to human calcitonin in patients resistant to salmon calcitonin. Lancet, 2, 1412, 1974.

32) Frost, H.M., The ADFR concept revisited. Calcif. Tissue Int., 36, 349, 1984.

33) Hesch, R.D., Heck, J., Delling, G., Keck, E., Reeve, J., Canzler, H., Schober, O., Harms, H., Rittinghaus, E.F., Results of a stimulatory therapy of low bone metabolism in osteoporosis with (1-38)hPTH and diphosphonate EHDP, Klin. Wschr., 66, 976, 1988.

34) Riggs, B.L., Nelson, K.I., Effect of long term treatment with calcitriol on calcium absorption and mineral metabolism in postmenopausal osteoporosis, J. Clin. Endocrinol. Metab., 61, 457, 1985.

35) Ott, S.M., Chesnut III, C.H., Calcitriol treatment is not effective in postmenopausal osteoporosis, Ann. Intern. Med., 110, 267, 1989.

36) Mc Carron, D.A., Morris, C.D., Blood pressure response to oral calcium in persons with mild to moderate hypertension, Ann. Intern. Med., 103, 825, 1985.

37) Lipkin, M., Newmark, H., Effect of added dietary calcium on colonic epithelial cell proliferation in subjects at high risk for familial colonic cancer, New Engl. J. Med., 313, 1381, 1985.

38) Garland, C.F., Comstock, G.W., Garland, F.C., Helsing, K.J., Shaw, E.K., Gorham, E.D., Serum 25-hydroxyvitamin D and colon cancer: eight-year prospective study, Lancet, II, 1176, 1989.

39) Hodsman, A.B., Effects of cyclical therapy for osteoporosis using an oral regimen of inorganic phosphate and sodium etidronate: a clinical and bone histomorphometric study, Bone and Mineral, 5, 201, 1989.

40) Valkema, R., Vismans, F.-J.F.E., Papapoulos, S.E., Pauwels, E.K.J., Bijvoet, O.L.M, Maintained improvement in calcium balance and bone mineral content in patients with osteoporosis treated with the bisphosphonate APD, Bone and Mineral, 5, 183, 1989.

41) Storm, T., Thamsborg, G., Steiniche, T., Genant, H.K., Sørensen, O.H., Effect of intermittent cyclical etidronate therapy on bone mass and fracture rate in women with postmenopausal osteoporosis, New Engl. J. Med., 322, 1265, 1990.

42) Watts, N.B., Harris, S.T., Genant, H.K., Wasnich, R.D., Miller, P.D., Jackson, R.D., Ligata, A.A., Ross, P., Woodson III, G.C., Yanover, M.J., Mysiw, W.J., Kohse, L., Rao, M.B., Steiger, P., Richmond, B., Chesnut III, C.H., Intermittent cyclical etidronate treatment of postmenopausal osteoporosis, New Engl. J. Med., 323, 73, 1990.

43) Fleisch, H., Bisphosphonate als Therapeutika - experimentelle Untersuchungen und klinische Anwendung. Therap. Umschau, 42, 366, 1985.

44) Holz, G., Delling, G., Ziegler, R., Etidronsäure-Therapie bei Morbus Paget des Skelettes, DMW, 51, 1954, 1984.

45) Khairi, M.R.A., Altman, R.D., DeRosa, G.P., Zimmermann, J., Schenk, R.K., Johnston, C.C., Sodium etidronate in the treatment of Paget's disease of bone. A study of long-term results. Ann. Intern. Med., 87, 656, 1977.

46) Van Holten-Verzantvoort, A.Th., Bijvoet, O.L.M., Cleton, F.J., Hermans, J., Kroon, H.M., Harinck, J.I.J., Vermey, P., Elte, J.W.F., Neijt, J.P., Beex, L.V.A.M., Blijham, G., Reduced morbidity from skeletal metastases in breast cancer patients during long-term bisphosphonate (APD) treatment, Lancet, 983, 1987.

47) Coleman, R.E., Pamidronate (APD) treatment of bone metastases from breast cancer, in The management of bone metastases and hypercalcaemia by osteoclast inhibition, Rubens, R.D., Ed., Hogrefe & Huber, Toronto Lewiston N.Y. Bern Göttingen Stuttgart, 1990, p. 52.

48) Delmas, P.D., Chapuy, M.C., Vignon, E., Charhon, S., Briancon, D., Alexandre, C., Edouard, C., Meunier, P.J., Long term effects of dichloromethylene diphosphonate in Paget's disease of bone, J. Clin. Endocrin. Metab., 54, 837, 1982.

49) Ziegler, R., Scharla, S.H., Treatment of tumor hypercalcemia with clodronate, in Bisphosphonates and Tumor Osteolysis, Brunner, K.W., Fleisch, H., Senn, H.-J., Eds., Springer, Berlin Heidelberg New York London Paris Tokyo, 1989, p. 46.

50) Franke, J., Runge, H., Osteoporose: Diagnose, Differentialdiagnose und Therapie unter Berücksichtigung der Natriumfluorid-Behandlung, VEB Volk und Gesundheit, Berlin, 1987.

51) Mamelle, N., Dusan, R., Martin, J.L., Prost, A., Meunier, P.J., Guillaume, M., Gaucher, A., Zeigler, G., Netter, P., Risk-benefit ratio of sodium fluoride treatment in primary vertebral osteoporosis, Lancet, 361, 1988.

52) Riggs, B.L., Hodgson, S.F., O'Fallon, W.M., Chao, E.Y.S., Wahner, H.W., Muhs, J.M., Cedel, S.L., Melton III, L.J., Effect of fluoride treatment on the fracture rate in postmenopausal women with osteoporosis, New Engl. J. Med., 322, 802, 1990.

53) Pak, C.Y.C., Sakhaee, K., Zerwekh, J.E., Parcel, C., Peterson, R., Johnson, K.: Safe and effective treatment of osteoporosis with intermittent slow release sodium fluoride: Augmentation of vertebral bone mass and inhibition of fractures. J. Clin. Endocrin. Metab., 68, 150, 1989.

54) Hedlund, L.R., Gallagher, J.C., Increased incidence of hip fracture in osteoporotic women treated with sodium fluoride. J. Bone Min. Res., 4, 223, 1989.

55) Fuchs, C., Haase, W., Maurer, H., Pharmakokinetik eines Fluorid-Kalzium-Präparates zur Behandlung der Osteoporose. Arzneim.-Forsch./Drug Res., 34 (II), 832, 1984.

Index

INDEX

A

B

C

D

E

F

G

H